Tips & Traps for Remodeling Your Bathroom

Tips & Traps for Remodeling Your Bathroom

R. Dodge Woodson

McGRAW-HILL

New York   Chicago   San Francisco   Lisbon   London
Madrid   Mexico City   Milan   New Delhi   San Juan
Seoul   Singapore   Sydney   Toronto

The McGraw·Hill Companies

Cataloging-in-Publication Data is on file with the Library of Congress

1 2 3 4 5 6 7 8 9 0   DOC/DOC   0 1 0 9 8 7 6 5

ISBN 0-07-145043-2

The sponsoring editor for this book was Cary Sullivan and the production supervisor was Pamela A. Pelton. It was set in Garamond by Lone Wolf Enterprises, Ltd. The art director for the cover was Handel Low.

Printed and bound by RR Donnelley.

Interior clipart images courtesy of www.clipart.com.

McGraw-Hill books are available at special quantity discounts to use as premiums and sales promotions, or for use in corporate training programs. For more information, please write to the Director of Special Sales, McGraw-Hill Professional, Two Penn Plaza, New York, NY 10121-2298. Or contact your local bookstore.

 This book is printed on recycled, acid-free paper containing a minimum of 50% recycled, de-inked fiber.

Dedication

I dedicate this book to Adam and Afton, the two best children a father could ask for. Afton has supported my writing since my first book, and Adam has never complained about my need to finish a chapter before venturing into our woods. They truly are the best children I could have dreamed of.

Contents

About the Author

R. Dodge Woodson has been involved in the building trades for 30 years and has been a self-employed contractor for 25 years. He is the owner of The Masters Group, Inc., a general contracting, remodeling, and plumbing firm in Maine. Woodson has written dozens of books on the industry for both consumers and professionals.

Introduction

Are you thinking of hiring a general contractor for home improvements or remodeling? Has the thought of saving thousands of dollars by being your own general contractor crossed your mind? Most homeowners seeking to improve their homes either hire a general contractor or act as their own construction manager while hiring subcontractors. In either case, this book is one of the most important tools that will be found on the job site.

Adding space to your home or improving existing living conditions can be a very traumatic time. But, it doesn't have to be. With the right knowledge, you can maintain control of your job. It will be easier on you to hire a general contractor, but there is a lot of money to be saved if you act as your own general contractor.

Almost anyone researching the rules of the road for remodeling has discovered horror stories about doing business with contractors and subcontractors. These stories are true. R. Dodge Woodson, the author, has been in the business for 30 years. He shares many of his own experiences on these pages. Best of all, he tells readers what to watch out for and how to avoid costly mistakes before, during, and after a home improvement or remodeling job.

Woodson has compiled a career of information here to help and protect you. For the mere cost of this book, you may save thousands of dollars on your job. Even more important, it is likely that you will avert disaster by not making the types of mistakes that many homeowners and inexperienced general contractors make.

Thumb through these pages. Notice the bullet lists, the tip boxes, and the numerous sample forms. The author has taken a serious, complicated subject and turned it into an accessible, easy-to-understand guide for homeowners. The writing is concise and the illustrations point out key factors toward a successful job.

You don't have to be a victim of unscrupulous contractors. Woodson will show you how to avoid them. Additionally, you will learn how to manage reputable contractors and assure yourself of quality work that comes in on budget and on time. Your home may be your single largest investment; don't risk it to renegade contractors. Learn how to protect yourself, your finances, and your home with this reader-friendly roadmap to success.

Go ahead and spend a little time looking over the chapters. It will not take long to see the value of Woodson's invaluable experience and advice. You don't have to go it alone. Take the words of a veteran contractor with you along every step of your remodeling adventure. Pick and choose the topics that you need, but don't go home without this essential element of your new project.

1

Planning Your Job

The planning phase of your bathroom-remodeling project is very important. In fact, proper planning is the key to success in any remodeling venture. It is during this phase that the mold is cast for your entire job. If a mistake is made in the planning phase, the result can be both dissatisfying and costly.

Why is planning so important? Planning is critical to a successful remodeling job for several reasons. It allows you to establish a budget, a timetable, and a goal for the desired results. For example, if you are working on a limited budget, as most homeowners are, it will be necessary to separate your needs from your desires.

Defining an accurate timetable can be crucial in bathroom remodeling. If you have only one bathroom in your home, knowing how long it will be out of commission will enable you to make alternate arrangements.

A PLAN

Before you begin to tear out your existing bathroom, it helps to have a plan for its replacement. If you go into a major remodeling job without a plan for the finished product, you are not likely to enjoy the experience. Can you imagine ripping out your bathtub only to find out that the replacement tub you want, that you thought could be picked up at the store, will really take six weeks to get? Could you live with a dysfunctional bathroom for six weeks? You could, but it wouldn't be fun.

Planning a major remodeling job is not something you do on a Saturday afternoon; proper planning takes time and effort. There will be phone calls to make, specifications to draft, and much more. What else will you need to do in the planning phase? This chapter is going to show you step-by-step what to do and how to do it. Once you have finished this chapter, you will be much better prepared to plan your job.

Bathrooms are one of the best rooms in a home to remodel when you are thinking of recovering the cost of your investment.

NEEDS AND DESIRES

Remodeling jobs evolve from either needs or desires, and you should identify the difference between them before you make financial commitments. If you are planning to remodel your bathroom, you are fortunate in your choice of the room to remodel. The kitchen is normally considered the most important room in a home to remodel, and bathrooms are considered the second best choice. Both bathrooms and kitchens are fine projects to invest your money in.

If you review statistics on which types of remodeling jobs are most likely to pay for themselves when a home is sold, you will see that kitchens and bathrooms normally control the first two spots on the charts. This knowledge is comforting, but don't allow it to lull you into a false sense of security. While it

Figure 1.1   Bathrooms can be as plain or as fancy as your budget allows. *Courtesy of Wellborn Cabinet, Inc.*

is true that bathrooms and kitchens are great places to invest your improvement money, you must invest wisely.

Bathrooms are expected to have certain components. They are expected to have toilets, lavatories, and bathing units. These are the most basic essentials, and there are many more suitable elements that could be included in a bathroom. There are also a host of add-on products available that are not mandatory equipment. To cull the crop of possibilities, you must separate your needs from your desires. To understand the difference between a need and a

> Investing too much in any project can result in lost money, and so can installing non-conforming products and materials.

Figure 1.2   A raised tub and lots of windows create a feeling of space and luxury. *Courtesy of Armstrong*

desire, let's look at an example for a typical bathroom-remodeling job.

Assume that you have a standard bathroom that you want to update. It has a wall-hung lavatory, a two-piece toilet, and a bathtub with no shower facilities. The existing floor structure is in good condition, but the old vinyl flooring is worn and dull. The walls and ceiling are painted drywall, and there is an old metal medicine cabinet with a mirror over the lavatory. The room is functional but unpleasing to the eye. What will you do to give this room a facelift?

If you want to make a safe investment, you will not get carried away with fancy options; you will make improvements,

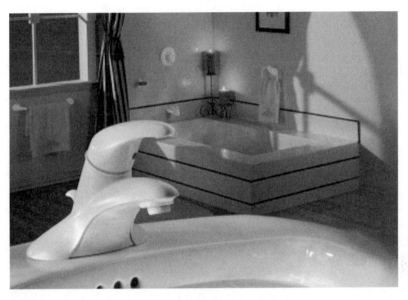

Figure 1.3   Fixtures such as this faucet can be an important element in your decorating plan. *Photo courtesy of Moen, Inc.*

but they will be simple and relatively inexpensive. For example, you will replace the toilet with a new two-piece model. The wall-hung lavatory will be replaced with a modest vanity and cultured marble top. New faucets will be installed for the lavatory and bathing unit, and the old tub will be replaced with a sectional unit that allows for use as both tub and shower. You will repaint the walls and ceiling and replace the old floor covering with new vinyl flooring. The ugly metal medicine cabinet will be replaced with an attractive oak cabinet and mirror. The fluorescent light that was over the old medicine cabinet will be replaced with a three-light oak strip.

If you follow the above procedures, you could hardly go wrong. The bathroom would be modernized and functional without being luxurious or expensive. This type of remodeling would meet all common needs, and it would be likely to return most of its cost when you sell your house. But is this what you want to do?

Figure 1.4   Traditional style blends with modern fixtures. *Photo courtesy of Moen, Inc.*

Suppose you want to replace the old tub with a whirlpool bath; how would that affect your job? Unless you have health problems, the whirlpool must be considered a desire, not a need. The extra expense of the whirlpool might be recoverable when you sell your house, but not necessarily. Installing such a bathing unit would extend the risk you are taking financially.

Perhaps you have always wanted a one-piece toilet and pedestal lavatory in your bathroom; would this be the time to install them? It could be, but you must acknowledge the fact that these items are desires, not needs. The bathroom will function just as well with a simple lavatory and toilet as it will with fancy fixtures. The extra money, a considerable amount, that you spend on a pedestal lavatory and one-piece toilet may not add to the value of your bathroom. The added expense is fine as long as you recognize that the money may not be recovered and that you are buying items you want, not items you need.

This type of needs-versus-desires evaluation will enable you to realize your goals without busting your budget. It is easy to become enthralled with the excitement of getting a new bathroom, but you must temper the excitement with logic and sound judgment. While many upgrades, such as wallpaper and tile, may be fine for your project, don't allow your dreams to exceed your financial capabilities and expectations.

BUILDING A VIABLE BUDGET

Building a viable budget for a remodeling project is not always easy. Every time you visit a store or thumb through a remodeling magazine, your wish list can grow. Unless money is no object in your project, self-discipline must prevail over sudden emotions. Building a budget will take time and effort, but it will also protect you from financial disasters.

Don't begin any work until you have a viable budget and cost projections that fall within that budget. Starting a job without this information is likely to result in a very expensive lesson that you would prefer not to learn from firsthand experience.

How Will a Budget Protect You?

How will a budget protect you? A budget, broken down by categories, will define your parameters for each type of expense involved with your job. For example, if you have budgeted $150 for a lavatory top, you will not, or at least you should not, spend $500 for a pedestal lavatory in a high-fashion color. Setting up your budget in phases will help keep your spending patterns on track.

How Will You Develop Your Budget?

How will you develop your budget? A good budget will start at the beginning and include every phase of work to be done. It will be broken down into several categories, and it will be laid out to show projected costs and actual costs.

The more detailed the budget breakdowns are, the better off you will be. For example, you could have a broad category, such as plumbing, or you could break it down into smaller sections. In the case of plumbing, you might have a subcategory for labor, fixtures, and rough materials. This would be helpful if you are doing your own plumbing, but if you are hiring a contractor, the budget breakdown might be different.

It is wise to include a budget amount for each plumbing fixture. Looking at lump-sum figures is not always as graphic as seeing the item-by-item costs. If you see quotes from three plumbers for $3,500, you might mumble about the high cost but not know if the price could be reduced. When you can see that you are paying $125 for a toilet, $400 for a bathtub, and $300 for a vanity and top, you can begin to evaluate the options you have. Seeing individual prices will allow you to alter your plans to maintain your budget.

When hiring a contractor, you should work with quoted prices for the services and materials to be provided. Under these circumstances, the breakdown between labor and rough materials is not very important. You are able to work with firm quotes and contracts that should detail all specifications for the work being done.

If you don't establish a viable budget, your job's cost could get out of hand and result in an unpleasant experience. Before you run all over town writing checks and charging materials, you should know, within a reasonable variance, what the total of your expenditures will be.

FINANCING THE WORK

Bathroom remodeling can get very expensive, and for many people this necessitates financing the work. While some people detest the idea of financing anything, some types of home-improvement financing may work to your advantage.

Equity Loans

Getting an equity loan on your home to pay for the work could result in some tax advantages for you. Not only is it possible that the interest paid on this type of loan is tax- deductible, but it is also a sure bet that you will still have your cash. If you can structure financing that offers tax advantages and invest your cash wisely in other areas, you may come out ahead of the game.

In-Home Financing

Many contractors have relationships with lenders that allow you to take advantage of in-home financing. The rates and terms of this type of financing are usually not as good as what you could obtain from your own bank, but there are times when this type of loan may be justified. If you decide to accept in-home financing, make sure that you understand all the terms and conditions. You should have an attorney review the paperwork and render an opinion on it before signing on the dotted line.

PLANS AND SPECIFICATIONS

While simple remodeling will not require the use of complex blueprints, it is necessary to have a set of plans, even if they are only simple line drawings and specifications, for the work you want done. This is especially important if contractors will be doing the work for you.

It is impossible to obtain accurate bids from contractors unless they are provided with identical plans and specifications. If you don't provide detailed plans and specifications, the contractors may all bid the job based on different

> Require all contractors bidding your work to provide a detailed outline of the work and products that they will be providing you for a set price. Make sure that all contractors have identical plans and specifications to work with when working up your quotes.

interpretations; the result will be bids that cannot be compared fairly.

Some people don't have the artistic ability to draw working plans, but all homeowners can create detailed specifications for their jobs. You can have professionals draw your plans for you, but you must create your own specifications; nobody else knows what you want. This part of your planning is essential, and it should be done before bids for labor and materials are sought.

PLANNING FOR INCONVENIENCES

Planning for inconveniences is a part of any remodeling job, but it is especially applicable to kitchens and bathrooms. If you were having your basement converted to finished living space, you would have to put up with noise, dust, and the traffic of contractors, but the level of inconvenience with kitchen and bath remodeling exceeds that of most other forms of remodeling.

When your kitchen is torn apart and reduced to a subfloor and bare walls, it can make life a little awkward. If the only toilet in your house has been removed to allow the installation of new floor coverings, you may wish you knew your neighbors better. The sudden need for toilet facilities can make you very aware of how much you depend on modern conveniences.

To avoid going hungry or begging your neighbor for the use of a bathroom, you should plan for known inconveniences. If your plans call for a total bathroom renovation, it is logical to assume that you may be without bathing facilities for a few days. Your toilet could be out of service for several hours at a time, and your lavatory may not be back in working order for days.

Before you begin your remodeling project, sit down and draw a mental picture of how the job will go. If you don't know enough about what will be done to create a plausible picture, ask your contractors to explain the sequence of events. If necessary, outline the work on a piece of paper; this may help you to see and avoid complications before they happen.

For example, assume the electrical wiring in your bathroom is going to be updated. What will happen if you decide to add circuits, such as a circuit for a whirlpool tub?

There is much more to major remodeling jobs than what first meets the eye. Look ahead, plan carefully, and be prepared for unexpected changes in your plans. If you do this, you will avoid many problems and be better prepared for trouble that cannot be dodged.

SETTING A REALISTIC TIMETABLE

Another part of your planning should include setting a realistic timetable for the completion of your job. This is often easier said than done. Remodeling jobs rarely go as planned, and staying on schedule can be very difficult. If one aspect of the job gets out of sync, the whole job can be affected. For example, if the lavatory top is not delivered on time, the plumber will not be able to install the lavatory. This delay can have a ripple effect with other trades, such as the electrician. If a cabinet delivery is delayed, the whole job can come to a halt. There are many small problems that can bring a big job to a screeching halt.

There are many circumstances that no individual can control fully, and this is why organizing a completion schedule that will work is not easy. The best you can do is to create a schedule and work hard to maintain it. This may mean hounding contractors or suppliers to do what they have promised, but in all cases staying on schedule will take effort. Jobs don't run themselves; if they did, there would be no need for general contractors.

Meeting deadlines on a production schedule can be achieved if the deadlines are realistic. If you allow adequate time for each phase of work, the chances are good that you can meet your self-imposed deadlines. There will, of course, be times when you can't control delays. Invariably, some jobs are plagued with bad luck; the wrong cabinets will be shipped, the bathtub will be damaged, the painter won't show up for days, and so on. Is it bad luck or bad planning? Usually it is a deficiency in organizational skills and follow-up effort, but there are times when the best efforts cannot alleviate the problems.

How Can You Set Realistic Time Goals?

How can you set realistic time goals? If you are hiring contractors to do your job, ask them how long their portions of the job will take. Consult cost-estimating manuals that give estimates for the time needed to complete various phases of work. Compare the data in the estimating manual with the answers given by your contractors. If the two estimates are similar, you should be right on track. When the time estimates differ substantially, investigate further. For example, if you are examining three electrical bids and one electrician says the job will take two days, ask the other electricians how long they believe the work will take. Once you have their time projections you can begin to answer your question about the time required to complete the electrical phase of the job.

> While you cannot control all aspects of your job, there is much you can do to keep it on schedule. For example, if you inspect all materials as soon as they are delivered, you can avoid some loss of time.

A form can be used to list each phase of work to be done and the estimated time allowed for the work. If you are going to do the work yourself, setting a completion date will be a little more difficult. You can still use cost-estimating manuals to help determine your time needs. Many of these guides list the number of hours a job should take to complete when professionals are doing the work. Some of the guides give advice on how to adjust the time estimates to allow for a lack of professional experience.

Remodeling jobs almost always take longer to complete than anyone except seasoned remodelers expect. Once you have outlined a production schedule, build in some extra time for unexpected work and delays. For example, when you remove the floor covering in your bathroom, you may find that the subfloor and floor joists have been damaged by a water leak at the base of the toilet. This type of unexpected work will increase the cost of your job and the time it takes to complete it. By building in a buffer for unexpected problems and mistakes in time estimates, you will be more likely to finish on or ahead of schedule.

CONTRACTOR CONSIDERATIONS

Contractor considerations may also play a significant role in the success of your remodeling project. Choosing the right contractors is not always easy, but it is necessary. The wrong contractors can turn your remodeling dream into a nightmare. Here are some tips for hiring contractors:

- Any contractor you hire should be properly licensed and insured.
- You may find it beneficial to work with bonded contractors.
- Don't take chances with part-timers who are not licensed or insured.
- Make sure your contractors obtain all needed permits and inspections.
- If you are going to hire contractors, choose them carefully.
- Check the references of potential contractors.
- Check with local agencies that report complaints against contractors.
- Don't give contractors large deposits for the work to be done.

THE END RESULT

Knowing what you want the end result of your remodeling to be is also important. Are you doing the job to build equity in your home or to satisfy your personal preferences? It is possible to do both at the same time, but the two don't necessarily

Cost Projections

Item/Phase	Labor	Material	Total
Plans			
Specifications			
Permits			
Trash container deposit			
Trash container delivery			
Demolition			
Dump fees			
Rough plumbing			
Rough electrical			
Rough heating/ac			
Subfloor			
Insulation			
Drywall			
Ceramic tile			
Linen closet			
Baseboard trim			
Window trim			
Door trim			
Paint/wallpaper			
Underlayment			
Finish floor covering			
Linen closet shelves			
Closet door & hardware			
Main door hardware			
Wall cabinets			
Base cabinets			
Countertops			
Plumbing fixtures			
Trim plumbing material			
Final plumbing			
Shower enclosure			
Subtotal			

Figure 1.5   Example of a cost projections form.

go hand in hand. If you are hoping to build equity in your home through the remodeling of a kitchen or bathroom, you must be selective in the work you do.

Spending too much on improvements can negate any equity you hoped to build. On the other hand, if you are making the changes to suit your personal desires and you are

Item/Phase	Labor	Material	Total
Light fixtures			
Trim electrical material			
Final electrical			
Trim heating/ac material			
Final heating/ac			
Bathroom accessories			
Clean up			
Trash container removal			
Window treatments			
Personal touches			
Financing expenses			
Miscellaneous expenses			
Unexpected expenses			
Margin of error			
Subtotal from first page			
Total estimated expense			

Figure 1.5 (*continued*)   Example of a cost projections form.

not worried about recovering your investment, as long as you can pay for it you can do it.

Take some time to decide what you hope to gain from your remodeling and ask yourself some questions, such as:

- Do you want more light in your bathroom?
- Must you have a whirlpool tub to make your life complete?
- Will doing the work yourself and building several thousand dollars of equity in your home make you happy?

These are the types of questions to ask yourself. Allow adequate time to find the real reasons for your urge to remodel. If you dedicate enough time to the planning of your project, you are much more likely to wind up with results that will please you.

2

Drawing Your Own
Rough Plans

When you are planning a remodeling job, sketching your intentions is a good way to design a rough plan of action. Having a sketch of what your want to do will make planning the job much easier, and you don't have to be an architect to draw your own preliminary plans.

Line drawings are often very effective for interior remodeling jobs. If you are not making structural changes in your home, a simple line drawing may be all you need to get the job done. Even if you have no artistic ability or inclination, you can do a pretty fair job of sketching a rough plan with the help of a ruler and some graph paper.

If your job is complex enough to warrant elements such as cross-sections and elevations that you are unable to draw, there are many options for you to consider. Architects are one option, but plans drawn by architects are usually very expensive, and this can be cost-prohibitive in many remodeling jobs.

Let's imagine that we are going to remodel a bathroom but are not sure what we want it to look like. Where should we start our planning? It would be nice if we could look through a book of floor plans, but finding a book that contains only bathroom plans is not easy. There are, however, many books available that contain house plans, and these books can be a good starting point.

BOOKS WITH HOUSE PLANS

Books with house plans also contain bathroom plans. The bathroom plans are not drawn individually; they are a part of the overall house plan, but that's all right. In the planning stage of a remodeling job, you are likely to be looking for ideas, and these books of plans can give your enough ideas to keep you busy all winter. Soon your problem will not be coming up with ideas; it will be deciding on which ideas to incorporate into your personal plans.

A single book of house plans may very well contain over a hundred different bathroom layouts. When you consider how many different books of plans are available, you might find thousands of designs to work with. Many of the designs will be similar, but each will have its own special features. You can borrow from several plans to come up with an ideal plan for your house.

House plans on the Internet are a good source of ideas for the remodeling of your bathroom.

When you look at the floor plans for various bathrooms, notice how simple they are. Most of them will consist only of simple lines. Some may be drawn in a dimensional perspective, but most will just be simple line drawings. Don't you think you could draw similar plans with the help of a few drafting tools and some graph paper?

SKETCHING YOUR OWN PLANS

Sketching your own plans can be very simple. If you gather a few basic drafting instruments, you can make the drawings look like those in the books of plans. There are templates available that enable you to draw sinks, toilets, and doors with ease; all you have to do is follow the stencil and you can't go wrong. An architectural-scale ruler is not too expensive, and it makes scaling a drawing very easy. If you don't want to buy a scale rule, you can use any ruler to create your own scale drawing. The grids on graph paper make scaling a rough drawing possible for anyone. The point is that you can draw your own preliminary plans.

Many software manufacturers sell consumer-grade drafting programs for computer-aided drafting. These programs are available on the Internet, in computer stores, and in large chain stores. Prices for these programs are often less than $50.

Templates

Inexpensive templates are available for every symbol you see on an average floor plan. Some of these symbols represent the following:

- Windows
- Doors
- Lavatories
- Toilets
- Bathtubs
- Other items

All of these items can be drawn with plastic guides. All you have to do is hold the template on the paper and trace a pencil around the stencils. Your results should look as good as those of a professional.

Scale Rulers

Scale rulers don't cost much, but they make drawing scaled drawings much easier than it would be with a standard ruler. If you want one of these devices, get an architectural-scale ruler; it will be of the most use to you for construction blueprints.

Scale rulers are equipped to handle several different scales. It won't matter whether you are working with a ¼-inch scale or a ½-inch scale; your scale ruler will convert the scale to real-world measurements. All you do is position the ruler on a line and measure it. If you are working with a ¼-inch scale, a 3-inch line will read as 12 feet on the scale ruler.

Regular rulers can be used to work with scale drawings, but you will have to do the math conversions on your own. This not only consumes more time, but it also makes mistakes more likely.

Graph Paper

Graph paper is the best type of paper to use for your preliminary drawing. The paper has grids spaced at proportional distances. You can assign each of these grids a distance value. For example, you might say the distance between horizontal lines equals 1 foot or 1 inch, depending on what you are drawing.

Draw all your plans to scale. I can remember numerous jobs where customers had wonderful designs that simply would not work since the drawings were not to scale. It's easy to get a large whirlpool tub between a cabinet and a wall if you don't scale the drawing, but when it comes time for the real installation, the appliance may not fit.

Once you have graph paper, a ruler, and a pencil, you can draw a rough plan of what you want to do. Having a template for your symbols will make the drawing look more professional, but you can get by without the template. Equipped with these tools, it takes almost no artistic ability to draw a simple floor plan.

GETTING PROFESSIONAL HELP

Getting professional help with your working drawings is not difficult, but it can be expensive. Architects are very well qualified for drawing your plans, but their fees are generally cost-prohibitive for simple home remodeling. Who can you turn to if you don't want to pay the price for an architect? You could check into drafting companies.

Drafting Firms

Many drafting firms draw blueprints and floor plans. Again, if you only need a floor plan, you can do it yourself, but if you are doing extensive work or structural work, you would be better off with professionally drawn plans. Most drafting companies will draw working plans from your rough drawing for very reasonable prices.

Free Drafting Services

There is another option that may not cost you anything. Some building-supply stores will provide free drafting services for their customers. If you are willing to buy your materials from the supplier, there is a good chance that you can get your plans drawn for free or for a very modest fee. However, the free plans may not be the bargain they appear to be. Before you commit to a deal, check the supplier's prices and the quality of the materials. It is possible that the price you will pay for the materials is far too much. If this is the case, you may be better off to pay someone else to draw your plans, so you can buy your materials wherever you want.

> College students are another possible source for low-cost plans. Students who are taking drafting classes may have the ability to give you good working plans for a low price.

HOW ACCURATE DO PRELIMINARY SKETCHES HAVE TO BE?

How accurate do preliminary sketches have to be? There is no rule that says preliminary sketches must be to scale, but if they

are not, it is easy to lose perspective on the job. It is not important for the symbols to be exact drafting symbols, but you
should strive to maintain a consistent scale, regardless of what
the scale is.

If you draw the floor plan without using a scale, objects
may appear much larger or smaller than they actually are.
Many homeowners sit down with pencil and paper and draw a
bathroom plan that looks spacious, but if the sketch is not
drawn to scale, there is no way to judge how spacious the
room will be. If you are freehand-drawing a bathtub, it is easy
to draw it to fit any space you want, but in real life you will
need a space with a width of 5 feet to install the tub. Many
homeowners fail to realize how large vanities, bathtubs, linen
closets, and similar items are. This distorts the options available
in a given space. To avoid disappointment when the construction starts, draw your sketches to scale.

JUDGING SIZES

Judging sizes for some items can be difficult for homeowners.
Take a look at the questions and answers below to get a feel
of the types of sizes that you may need to deal with:

- Q: Do you know how wide a typical vanity cabinet is?

- A: It is normally 24 inches to 48 inches feet wide.

- Q: What are standard widths for wall cabinets that are
  not custom-made?

- A: The most common widths for stock wall cabinets are
  12, 15, 18, 24, 30, and 36 inches.

- Q: How deep are most of these cabinets?

- A: The depth of most wall cabinets is 30 inches.

- Q: If you are remodeling a powder room and need a small
  vanity, what might be the smallest stock size available?

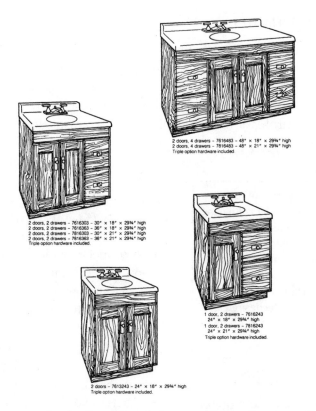

2 doors, 4 drawers – 7616483 – 48″ × 18″ × 29¾″ high
2 doors, 4 drawers – 7816483 – 48″ × 21″ × 29¾″ high
Triple option hardware included.

2 doors, 2 drawers – 7616303 – 30″ × 18″ × 29¾″ high
2 doors, 2 drawers – 7616363 – 36″ × 18″ × 29¾″ high
2 doors, 2 drawers – 7816303 – 30″ × 21″ × 29¾″ high
2 doors, 2 drawers – 7816363 – 36″ × 21″ × 29¾″ high
Triple option hardware included.

1 door, 2 drawers – 7616243
24″ × 18″ × 29¾″ high
1 door, 2 drawers – 7816243
24″ × 21″ × 29¾″ high
Triple option hardware included.

2 doors – 7613243 – 24″ × 18″ × 29¾″ high
Triple option hardware included.

Figure 2.1   A selection of vanity base cabinets. *Courtesy of Universal Rundel*

- A: A vanity with dimensions of 16 x 18 inches shouldn't be any problem to find.

Knowing these typical sizes will be important when designing your new bathroom.

How will you know what sizes to use for various items? You can look in catalogs for sizes, or you could go to building-supply centers and measure various items. Sizes are easy to come by.

> When you need sizes and rough-in dimensions, go on the Internet. Get to the web sites of manufacturers that make the products you are interested in. Many of these sites will provide dimensional information.

CODE REQUIREMENTS

Code requirements are another factor that you must consider when drawing your working plans. If you are only drawing a rough sketch and will have a professional prepare your working plans, you can get by without knowing code requirements; the professional will adjust your drawing to comply with them.

Many homeowners opt to save a few hundred dollars by not having professional blueprints prepared for their remodeling jobs. Sometimes these homeowners are able to communicate what they want to contractors and get it. But many times communication between the parties leaves something to be desired, and the job does not turn out as the homeowners had hoped.

Why do code requirements affect your drawing? They may influence the layout of your fixtures, outlets, and so on. For example, a toilet requires a clear space 30 inches wide for installation. The distance from the front of a toilet to another fixture is normally required to be at least 18 inches. If you were not aware of these types of requirements, you might lay out your bathroom in a way that would violate code requirements and have to be redesigned.

If you draw your own working plans and submit them for approval to the local code-enforcement office, the code officer will let you know if the drawing is in violation of the code. Door widths, ceiling heights, electrical outlets, and plumbing fixtures are where most of the spacing requirements will be scrutinized.

Once you have completed your preliminary plans, there are likely to be many changes made to them. It is best to make these changes before work is begun, but that is not always possible. While it is probable that there will be changes made after the job is started, it is best to firm up the plans as much as possible before starting work.

EXTENSIVE REMODELING

If you are planning an extensive remodeling job, you will benefit from having a good understanding of blueprints and design plans. While complex drawings are rarely used in minor remodeling jobs, they are common when doing larger jobs. For example, if you are expanding your kitchen and relocating your fixtures, appliances, and cabinets, you should have a detailed set of plans to work from. The need for solid working plans is essential if you want contractors to do the job the way you envision it. However, many bathroom-remodeling jobs do not require extensive drawings to get the job done.

It is easy for two people to see the same vision in a different light. For example, assume that you tell your plumber that you want an almond-colored, two-piece toilet. This seems like a clear description, doesn't it? Okay, now assume that you leave for work, and when you get home, the plumber has installed an almond-colored, two-piece toilet. You take one look at it and hate it. The toilet has a bowl that is molded with the contour of the integral trap, and you find the design to be repulsive. What are you going to do? The plumber provided what you asked for. It just isn't what you thought it would be.

If you are doing a major remodel, plans and specifications should be considered a compulsory expense. The money or time you spend in preparing precise plans and specifications can save you time, money, and aggravation.

You see, there are many styles of toilets, and some people prefer one style to another. While the toilet installed is a name brand and is commonly used, it does not look like what you had in mind. If you had given the plumber a make and model number to work with, you would have the toilet you wanted. As it is, you have a toilet you don't want, but you can't expect the plumber to replace it for free. After all, it does meet the

Figure 2.2   This bathroom has been carefully planned to maximize ease of use. *Courtesy of Armstrong*

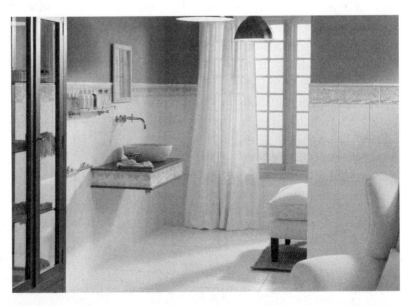

Figure 2.3   The unique layout of furniture and fixtures makes this bathroom exceptional. *Courtesy of Armstrong*

description you gave. This is why detailed specifications are needed on all jobs.

WHEN DO YOU NEED BLUEPRINTS?

When do you need blueprints? Blueprints, or at least line drawings, are beneficial on all jobs, but they are a near necessity for large jobs. Let's look at two types of jobs, one large and one small, and see how blueprints would affect them.

Minor Bath Remodeling

Minor bath remodeling does not require extensive blueprints. The job can be done without any drawings, but it may help to have a simple line drawing that depicts the finished product. Let's see how drawings might help under these circumstances.

Assume that you are not changing the location of any primary plumbing but you are replacing all your plumbing fixtures. If new fixtures will be installed in the exact locations of the existing fixtures, it should not be necessary to have blueprints to indicate fixture placement. Putting a new bathtub in the old opening and setting a new toilet on an existing flange do not require a lot of direction.

If all the work in the job will be this obvious, plans will not be needed, but good specifications will.

READING BLUEPRINTS

Reading blueprints is not difficult if you take your time and understand the symbols used on the drawings. In the case of most bath remodeling jobs, the blueprints will not be very complicated. However, it will still be necessary to understand how plans are drawn to scale and what the various symbols mean.

If you are working from a set of professionally drawn blueprints, everything you need to know to read them is likely to be included on the plans. There should be notes that indicate the scale used on the plans, and there should be a section that shows all the symbols and their meanings. This is called a legend.

Learning what the symbols and different types of lines represent is not difficult. Working with scaled drawings is easiest when you have a scale ruler, but any ruler will do. Most blueprints are drawn with a scale in which ¼ inch on the blueprint is equal to 1 foot in real life. In other words, a countertop that measures 2 inches on this type of blueprint would actually be 8 feet long.

> There are many times when detailed drawings can be avoided, but you should never hire contractors to do a job without explicit specifications.

It is important to note that not all blueprints are drawn to the same scale. In fact, it is possible for the scale to change from one page of the prints to the next. The scale often changes for cross-section details and elevations. Before relying on scale measurements, check each section of the plans for the scale being used.

If you are doing the work yourself, professionally drawn blueprints will be your guide. By checking the prints, you can see what size lumber is required, how thick the countertop should be, how far below the ceiling the cabinets should hang, and where your fixtures should be placed. A good set of blueprints will leave nothing to the imagination.

SPECIFICATIONS

We have already seen some examples of how specifications can influence the progress of a job. If you are working as a do-it-yourselfer, professionally prepared specifications can answer many of your questions. While you may know that an underlayment is needed for your new tile floor, you may not know what size or type to use. Professional specifications will tell you what to use in all your remodeling tasks.

The Masters Group, Inc.
PMB # 300 13 Gurnet Road
Brunswick, Maine 04011
207-729-8357 (Phone)
207-798-5070 (Fax)
tmg1@mfx.net (Email)

Change Order

The Contract between the parties shall remain in full force and effect except as specifically amended by this change order.

1. **Parties To This Change Order:**

A. Contractor   The Masters Group, Inc.   207-729-8357
                       PMB # 300 13 Gurnet Rd.
                       Brunswick, Maine 04011

B: Homeowner or Lessee:_____
                       Name   Phone

                       _____
                       Address

2. **Date Original Contract Signed:** _____

3. **Changes in the Work Originally Contracted For:** _____

_____

_____

4. **Price Change:**

A: Original Contract Price: $_____

B: Revised Contract Price: $_____

5. **Acceptance of Change Order:**

Signature: _____
           Homeowner or Lessee   Date

Signature: _____
           The Masters Group, Inc.   Date

Each party must receive a copy of this signed contract before work can be started.

Figure 2.4   Change Order

If you are thinking that you don't need clear specifications when doing the work yourself, you are likely to be wrong. Even professional tradespeople rely on specifications drafted by architects and engineers. It is one thing to know how to install a floor joist and quite another thing to know which size joist to install. Every job benefits from clear specifications, regardless of who is doing the work.

If you provide a carpenter with a detailed set of plans and specifications and the job doesn't turn out right, you have grounds for insisting that changes be made, at no cost to you, to bring the job into compliance with the plans and specifications. You should have all contractors sign a copy of the plans and specifications to prove that they were given a set; this can be of great benefit if you are one of the unfortunate few who winds up in court with a cantankerous contractor.

> When you hire others to do remodeling for you, detailed specifications are the only way to ensure that the job will be done the way you want it done.

THE KEY TO UNDERSTANDING BLUEPRINTS AND DESIGN PLANS

The key to understanding blueprints and design plans is patience. Take your time in looking the plans over and familiarize yourself with the symbols and scales. Practice scaling distances on lines where the distances are given. If you see a wall with a noted length of 8 feet, scale it and see if you come up with the right measurement. Always confirm the scale on the section of drawing that you are working with. Look at the legend of symbols and search the plans until you can find and identify each one. Practice does make perfect when it comes to reading blueprints.

3

Solidifying Plans and Estimating Job Costs

S olidifying your plans before beginning any remodeling work will save time and money. Remodeling is known for its unexpected changes in plans, but you should do your best to avoid in-progress changes. Even after doing your best, you will probably experience some problems. Professional remodelers have problems with most jobs, so it is unlikely that you will avoid them. However, you can hedge your odds.

What can you do to solidify your plans? There are many ways to reduce the risks of on-the-job problems. If you will be using contractors on your job, you should meet with each of them prior to starting work. One common problem on many jobs is conflict between the different trades. It is not unusual for plumbers and heating mechanics to get in each other's way. Electricians sometimes block the paths of other trades, and painters and drywall finishers often argue about who is at fault for less-than-perfect wall finishes. Of course, if you are doing all the work yourself, there won't be anyone else to worry about. All you will have to do is stay out of your own way.

Most homeowners don't possess all the skills necessary to do a full-scale remodeling job without some professional help. However, any homeowner can gain enough knowledge to act as a general contractor, and this effort can save a considerable amount of money on the cost of a job. Just by acting as your own general contractor you can save between 10 and 30 percent on the retail price of a job. On a $12,000 bathroom remodel, this can amount to some serious money.

The minute you decide to involve other people in your job, you must make a commitment to refine your plans and specifications before starting any work. To show the importance of solid plans, let's look at a sample job and how it might be affected by a lack of proper preparation.

A MAJOR BATHROOM REMODEL

A major bathroom remodel can be a big undertaking. The work required in this type of job can involve a multitude of trades. What trades are likely to be involved? The work may involve any of the following:

- Demolition
- Rough carpentry
- Trim carpentry
- Plumbing
- Electrical
- Heating
- Drywall installation
- Drywall finishing
- Painting
- Floor covering
- Insulation installation
- Tile installation

Bathroom Remodeling Notes

Customer's Name: _____

Customer's Address: _____

Job Address: _____

Customer's Phone Number: _____

What type of space is below the bathroom? Is it a basement or other living space?
_____

What type of space is above the bathroom? It is attic space or living space?
_____

How wide is the bathroom door? _____

What type of door is installed on the bathroom opening? _____

Does the bathroom have a window that opens? _____

Is there an exhaust fan in the bathroom ceiling? _____

What type of flooring exists in the bathroom? _____

Does the subfloor seem solid, especially around the toilet and tub? _____

What are the dimensions of the floor? _____

What are the dimensions of the ceiling? _____

What are the dimensions of the wall areas? _____

How wide and deep is the shower or tub opening? _____

How far is the center of the toilet from the wall behind it? _____

What type of bathing unit exists in the bathroom? _____

Figure 3.1 Bathroom Remodeling Notes

Does the water pipe for the toilet come out of the wall or out of the floor? _____

Is there an access panel for the bathing unit? _____

Do the pipes for the lavatory come through the floor or the wall? _____

What type of drainpipe is existing? _____

What type of vanity top is wanted? _____

How wide will the vanity be? _____

What type of faucets are wanted? _____

What type of water pipe is existing? _____

Is there any exposed plumbing along the walls or baseboards? _____

Will plumbing fixtures be going back into the same locations, or will they be moved?

What type of roof does the house have? _____

Is there a GFI circuit for the bathroom? _____

What type of light fixtures are existing? _____

What type of heat is in the bathroom? _____

How far is the electrical panel from the bathroom? _____

What is the access for running wires to the panel? _____

What brand of circuit breakers are being used? _____

Is the electrical panel 100-amp or 200-amp? _____

Does the electrical box appear to have blank spaces available in it? _____

Does any of the heat need to be moved? _____

What type of walls are existing? _____

What type of ceiling is existing? _____

Figure 3.1 (*continued*)   Bathroom Remodeling Notes

Do the walls and floor seem level and plumb? _____

How difficult will the ripout be to get out of the house? _____

How is the access for hauling away debris? _____

What type of flooring is wanted? _____

What brand and type of plumbing fixtures are wanted? _____

Will a pedestal lavatory be used? _____

What types of walls will be used? _____

What type of ceiling will be used? _____

What type of cabinets will be needed? _____

What type of medicine cabinet or mirrors will be wanted? _____

Are tub/shower doors wanted, or is a curtain okay? _____

Is ceramic tile needed? _____

Are there any skylights to work around? _____

What type of accessories (toilet-paper holders, towel racks, etc) are needed?

What type of wall covering, such as paint, is wanted? _____

Are we doing anything with the door? _____

What type of trim will be installed? _____

What do we need to trim? _____

Are we painting or staining the trim? _____

NOTES:

_____

Figure 3.1 (continued)   Bathroom Remodeling Notes

- Cabinet installation

- Other types of work

Can you do all these jobs? If you can, you are one of only a few homeowners who can handle this type of job without professional help.

Since many homeowners will not feel comfortable doing any of their own work, let's look at this job through the eyes of a person who is acting as a general contractor. If you plan to do portions of your own work, you can substitute your labor for the trades you are comfortable doing. Let's start the job from the beginning and see what problems might arise.

The first step will be to rip out the existing fixtures and floor covering. This phase of the job doesn't require a lot of skill, but you must be careful not to damage primary systems. For example, you have to cut the water off to the plumbing fixtures before removing them, and caution must be exercised when removing light fixtures.

Considerable dust can infiltrate a home during the demolition phase of a remodeling job. Plan on sealing off the work area with plastic to contain the dust. Use masking tape to secure the plastic to door openings and weight the bottom of the plastic with a piece of wood to allow ingress and egress to the work area.

Once the rip-out is complete, you must get rid of the debris. Here are some questions to ask yourself at this phase of the job:

- Have you made plans for having the debris removed? If you haven't, this is your first problem.

- Should you have arranged for a temporary trash receptacle to place the debris in to be hauled away at a later date?

- Should you have laid the debris on a tarp in your yard and scheduled someone to pick it up and dispose of it?

- Can you haul the debris to a dump area yourself?

Don't wait until you have debris to figure this out. Make a plan in advance. The removal of debris can require time to arrange, so it is important to take care of this before demolition work is started.

What will you do now that the rip-out is complete? Normally, any needed alterations to the heating or plumbing systems will be done first. It is unlikely that there would be a conflict between these two phases in an average bathroom-remodeling job.

> Most rough-in work for the mechanical trades, which include plumbing, heating, and electrical work, requires an inspection from the local code-enforcement officer before the work is concealed. Don't allow this work to be hidden by insulation or wall coverings until you have proof of approved inspections.

After the plumbing and heating rough-ins are done, the electrical work is next. Again, in a typical bath remodel this shouldn't cause any conflict. However, if you were expanding the size of the bath or remodeling a kitchen, the plumbers, electricians, and heating mechanics might all get in one another's way.

When all the rough-ins are complete and inspected by the local code officer, you are ready for the insulation installation. There may be no need for additional insulation in an interior-remodeling job After any required insulation work and inspections, you are ready to hang drywall. This phase of the job should go smoothly. After the drywall is hung, you are ready to tape and finish it. This is a dusty job and it takes some special skills, but there should not be any special problems with this phase of the job.

Now you are ready for paint. What happens when the paint is applied and the finish of the walls is not acceptable? Is it the

> Some jurisdictions require insulation to be inspected prior to concealing it.

drywall contractor's fault or the painter's fault? This is a debate that can rage back and forth for what seems an eternity. There is one way to solve this problem, but you must plan for it in advance.

Insist that the drywall contractor apply a coat of primer to the walls before leaving the job. When the primer is applied, any defects in the finish work will show up. This pins the fault on the right party. If the walls pass the primer test, you can move on to the painter; if the walls don't look good after the painting, you can hold the painter responsible. After the paint is done, you are ready to install the flooring. It is not unusual for the flooring contractor to mar the finish of new walls, so watch this phase closely.

> Insist that the drywall contractor apply a coat of primer to the walls before leaving the job. When the primer is applied, any defects in the finish work will show up.

After the flooring is in, you are ready for the plumber to set fixtures. Then the heating mechanic can trim out the heating system and the electrician can finish the electrical work.

Trim carpentry can be done anytime after the finish flooring is in place, but it is often scheduled after the mechanical trades are complete.

When everything else is done, the painting contractor will probably have to come back to paint the trim and touch up any places blemished by the other trades.

When all the work is finished and inspected by the code officers, all that is left is the cleanup work. Once the cleaning is done, you have a fresh, revitalized bathroom. This doesn't sound too hard, does it? Well, it is not always as easy as it sounds. What could go wrong? Lots of problems could come up.

> Avoid sending multiple trades into the job on the same day. If you have more than one trade on the job at a time, you will not know whom to blame for damages that occur.

Suppose your flooring contractor doesn't show up when scheduled—what will that affect? It will affect your entire finish schedule. For example, you wouldn't want the plumber to install the toilet and vanity before the flooring goes in. It is very possible that you would have to reschedule all the trades that follow the flooring.

What would happen if one of the trades didn't pass the rough-in inspection? This problem could prohibit you from moving ahead with the drywall installation. If you have to postpone the drywall, the whole job slows down.

The list of potential problems could go on and on, but you should be getting the idea that loose ends can affect your production schedule. These risks escalate when the size of the job is larger, as with major kitchen remodeling.

HOW CAN YOU AVOID ON-THE-JOB PROBLEMS?

How can you avoid on-the-job-problems? How you avoid problems will depend largely on the type of job you are doing and the number of people involved. To illustrate this, review the following list of problems and solutions:

- The first step is to have all agreements between you and your contractors in writing. The value of well-written contracts is immense. A good contract can protect you and your home while giving you control over the contractors.

- Open communication between the trades is also instrumental in the completion of a successful job. Having everyone who will be involved on the project get together on the job before work is started can help to eliminate confusion and conflicts before they affect your production schedule.

- Material deliveries are a frequent source of on-the-job problems. It is not uncommon for deliveries to arrive days after they were scheduled to appear on the job. If you have a place to store materials, order early and check each delivery carefully.

- If you make any changes in your plans or specifications after the job is started, make sure that all trades are aware of the changes. While you might not think that changing from a flush-mount to a recessed medicine cabinet will affect anyone but the carpenters, the mechanical trades could be adversely affected by such a change. The recessed cabinet might conflict with the plumber's vent pipe, the heating mechanic's ductwork, or the electrician's wiring plan. Any change could affect multiple trades.

A smooth remodeling job requires good organizational skills and a team effort. If everyone is working individually, oblivious to what others are doing, problems are sure to arise.

Figure 3.2   A beautiful bathroom is the end result of good planning and organization. *Courtesy of Wellborn Cabinet, Inc.*

Figure 3.3   Since you must live with your decisions, the special features in this bathroom require careful planning. *Courtesy of Wellborn Cabinet, Inc.*

SOLID PLANS MAKE FOR SMOOTH JOBS

Solid plans make for smooth jobs. There will more than likely be some changes you cannot predict until the job is started, but make an effort to get all changes made before the work is in progress. No one likes to see a job lose its momentum. You will be stuck with a job that drags on and on, and the contractors who bid the job for a flat-rate fee will lose money.

I cannot stress enough how important preplanning is for a successful remodeling job. If you plan well in advance and remain organized at all times, you can handle the unexpected much better than many professional contractors do.

> Whenever there is a change in your plans or specifications, insist on written change orders that formalize the proposed changes. Get it in writing!

ESTIMATING YOUR COSTS

Estimating your costs for a big remodeling job may seem like a formidable task, but it doesn't have to be. Cost estimating is a vital part of any remodeling job, and it is a job in itself. There are people who get paid good salaries for doing nothing but cost estimating. With today's technology, many professionals rely on computers and software for their estimating needs. While it is unlikely that you have access to high-tech estimating programs, you needn't feel helpless. There are many effective ways for you to develop accurate estimates for your labor and material needs.

If you have the skills to do all your own work, estimating your costs will be a little easier; all you will have to be concerned with are the prices for your materials. However, if you are an average homeowner, you will have to rely on some professional help to realize your remodeling goal. This will necessitate estimating the cost for those professional fees.

Since most homeowners need professional help, we will explore the ways to estimate both material prices and professional labor costs. While any one of the methods can produce an accurate estimate, you should combine the methods to assure the most accuracy in your estimates.

ESTIMATING WITH THE HELP OF CONTRACTORS

Estimating with the help of contractors is the easiest way to figure the cost of your job. Essentially, all you have to do is to ask several contractors to give you bids for the job. The contractors will be glad to give you estimates or quotes for the prices of labor and materials to complete your project. This is a simple process, but there are some aspects of this type of estimating that can skew the numbers. If you want accurate figures, you should follow some basic rules. What are those rules? Let's find out.

Plans

If you are going to allow contractors to do your job for you, a set of plans for what you want done is an absolute necessity. Even if the job is a small one, you will want to give each contractor bidding the work a set of plans. Without the plans the contractors cannot possibly bid the job competitively.

The plans you issue to contractors don't have to be elaborate, but they must detail all important aspects of the job. A rough sketch of what you want done will probably be adequate, but don't attempt to estimate your job with contractors until you have some type of plans to hand out.

Specifications

Specifications are as important, if not more important, than plans when working with contractors. When you put your job out to bids, you want all contractors pricing the same materials and services. This is only possible with a set of plans and a detailed set of specifications.

Many contractors use the phrase "or equal" to build in flexibility for substitutions; don't allow such a clause in your quotes or contracts. To obtain accurate estimates, you must be certain all contractors are bidding the job identically.

It is not enough to list your specifications casually. You should not say that you want a new toilet, a new lavatory, and new faucets for the lavatory and bathtub. The specifications should include all technical data needed to identify the specific items you want. This information will normally include a model number, style, color, and similar information.

Substitutions

Sometimes contractors make substitutions in materials when bidding work. If you allow this to happen, it will be impossible to compare the bids you receive accurately. Many contractors

COST ESTIMATES FORM

Cost Projections For Bathroom Remodeling

Item/Phase	Labor	Material	Total
Plans			
Specifications			
Permits			
Trash container deposit			
Trash container delivery			
Demolition			
Dump fees			
Rough plumbing			
Rough electrical			
Rough heating/ac			
Subfloor			
Insulation			
Drywall			
Ceramic tile			
Linen closet			
Baseboard trim			
Window trim			
Door trim			
Paint/Wallpaper			
Underlayment			
Finish floor covering			
Linen closet shelves			

(continues)

Figure 3.4　Example of a cost estimates form

Item/Phase	Labor	Material	Total
Closet door & hardware			
Main door hardware			
Wall cabinets			
Base cabinets			
Counter tops			
Plumbing fixtures			
Trim plumbing material			
Final plumbing			
Shower enclosure			
Light fixtures			
Trim electrical material			
Final electrical			
Trim heating/ac material			
Final heating/ac			
Bathroom accessories			
Clean up			
Trash container removal			
Window treatments			
Personal touches			
Financing expenses			
Miscellaneous expenses			
Unexpected expenses			
Margin of error			
TOTAL ESTIMATED EXPENSE			

Figure 3.4 (*continued*)   Example of a cost estimates form

use the phrase "or equal" to build in flexibility for substitutions; don't allow such a clause in your quotes or contracts. To obtain accurate estimates, you must be certain all contractors are bidding the job identically.

Contractor's Qualifications

A contractor's qualifications may affect the price you are quoted for a job. When you are working towards a final decision on which contractor to select and which price is right, you should consider the contractor's qualifications.

A building contractor who normally builds new houses may not be well qualified to give accurate price estimates for remodeling. While this contractor may be very familiar with what it costs to build a house, remodeling may present circumstances that the builder is not experienced with in terms of pricing.

> When you are working towards a final decision on which contractor to select and which price is right, you should consider the contractor's qualifications.

A remodeling contractor who has not been in business long may not have the ability to produce accurate quotes. Some contractors will give a low price to get the work, but will they be around to finish the job? There are many situations in which a contractor's qualifications could affect the pricing and service you receive.

If you don't take the contractor's qualifications into consideration, you may base your plans on a given price and find out that the estimated price is far too low. You can avoid this problem by obtaining numerous bids for the job and comparing them. If a few are extremely low, avoid those contractors.

Time of Completion

The time of completion for a job can affect the price of the work. Some contractors will take a job at a lower price if they can use it as fill-in work. This allows the contractor to have something to work on when previously scheduled work does not go as planned. For example, if a contractor is scheduled to build a deck and it rains, what can be done to salvage the day? Since it is rarely productive to build decks in the rain, the contractor could come to your job and work inside.

When requesting bids for remodeling work, always get at least three bids. Having five bids is a better strategy.

If you are not in a hurry to get your job completed, this can be a good approach to take in getting a low price and a good job. However, if time is of the essence as it usually is with bathroom and kitchen remodeling, you will probably not be willing to allow the job to drag out for months. Before you accept what appears to be the best bid, establish the time of completion for your work.

Time of Day

What does the time of day have to do with your pricing estimates? It can have a lot to do with it. There are many licensed, insured, reputable contractors who work part-time. These contractors often have full-time jobs during the day and work their own business at night and on weekends. This group of contractors can offer attractive pricing.

If you don't mind having your nights and weekends interrupted with the inconveniences of remodeling, you can save some money by finding part-time contractors. However, if you don't want your hours off from work consumed with the noise, dust, and general commotion of remodeling, you will probably have to pay more for the work you want done. If you are shopping for the lowest price, find out why the lowest price is so low.

Quotes and Estimates

Do you know the difference between quotes and estimates? Estimates are just that, estimates. They don't guarantee a fixed price for a job. Since estimates are not quotes, it is hard to hold a contractor to an estimated price. When you are trying to estimate the cost of your remodeling project, you shouldn't base your figures on estimates provided by contractors. You need quotes.

Quotes are guaranteed prices that will not increase. They are usually good for thirty days. Once you make a commitment to sign a contract, the quoted price becomes the contract price and should not change. This is the only type of pricing that you can depend on.

Working with part-time contractors can be risky, but it can also be a great bargain. If the contractors are licensed and insured and can do the work within the time limits that you set, they could be a good deal. On the other hand, these contractors may not stay in business long enough to finish your job or to tend to any needed warranty work. You will have to decide between the risk and the reward.

Soft Costs

Soft costs are sometimes paid by contractors and sometimes paid by homeowners. When you are soliciting quotes for your job, you need to know if the quotes include all soft costs. What are soft costs? Soft costs are expenses such as permit fees, blueprints, and other fees that are not related directly to materials and labor. Before accepting any quote as the final figure for your job, identify all necessary soft costs and how they will be paid for.

MATERIAL TAKE-OFFS

What are material take-offs? They are simply lists of materials that will be needed to complete a job. If you plan to estimate your job without the help of contractors, it will be your responsibility to come up with a take-off of the materials you need. Why are these lists called take-offs? They are called take-offs because the lists are developed by taking material information off a set of plans. In other words, you would look at your plans and see that you need 120 square feet of underlayment to make a take-off for your kitchen floor.

Few homeowners will have the ability to make accurate take-offs from blueprints. For those homeowners working as do-it-yourselfers, this can be a problem. When you hire a plumber to replace the fixtures in your bathroom, you don't have to be concerned with how many compression ferrules or supply tubes will be needed. The plumber takes care of it for you. However, if you are going to replace your own fixtures, you will need to estimate the types and amounts of materials you will need. If you have never replaced plumbing fixtures before, you are likely to make mistakes in your estimates. Estimating the material needs for a complete bathroom remodel is much more difficult.

How will you make an accurate material take-off? You could guesstimate the needs for the job, but you will wind up with too many of some items and not enough of others. Since it is unlikely that you will be able to produce an accurate list of your needs from blueprints, the best option is to seek professional help.

What types of professionals will help you with your estimate for materials? If your plans were drawn by a professional, there is a good chance that that person will provide you with a list of materials needed to complete the work. You will have to pay for the information, but it should be accurate and the cost is not likely to break your budget.

Are there any options other than contractors and the professionals who draw plans for accurate take-offs? Yes—you can seek assistance from material suppliers. Almost any store that sells the supplies you need will be willing to help you establish a list of materials from your plans. Some places charge for this service, and others provide it free of charge so long as you buy your materials from them. Material suppliers can be of great help in estimating your costs.

ESTIMATING WITH THE HELP OF MATERIAL SUPPLIERS

Estimating your costs with the help of material suppliers is a fine way to establish realistic cost projections for the materials needed in your job. Not all take-offs from suppliers will be as accurate as those provided by an architect, but they will be close enough for most needs.

To get a supplier to estimate your material needs and costs, all you have to do is provide a set of plans and specifications. Asking a supplier to bid your job from plans and specs will not necessarily give you a detailed take-off, but it will give you a fixed price to work with. You should get prices from several suppliers and compare them.

Ask the suppliers to break their prices down into phases of work. For example, ask that you be given separate prices for cabinets, countertops, flooring, framing lumber, and so on. If you have several material bids that are broken down this way, you can compare each phase to spot mistakes the suppliers may have made. For example, one supplier may list eight base cabinets where another supplier showed nine. At first you may not know which supplier's take-off is correct, but you will know that someone is wrong, and that will enable you to establish your true needs.

Once you have obtained and reviewed several estimates for your materials, you will have a good idea as to what your costs will run. The other part of your job to estimate will be soft costs and labor. If you are doing the work yourself, estimating the cost of labor will not be important. The soft costs will probably include permit fees and working drawings, and you may not even need professionally drawn plans.

ESTIMATING WITH THE HELP OF COST-ESTIMATING GUIDES

Estimating with the help of cost-estimating guides is another way to project the expense of your job. With the use of these guides it is possible to predict the cost for both labor and materials. There are a number of cost-estimating guides available in bookstores, and most of them provide a lot of useful information.

There are drawbacks to depending on estimating books. These books can age and become out of date. Material prices can jump up and down frequently, and the books cannot foresee these changes. Since the books are written generically, the figures given may not be applicable in your particular situation.

Good estimating guides provide a way for you to convert generic estimates into regional estimates. Since prices in California are not the same as prices in Florida, the broad-brush estimates must be refined. Multipliers are normally used to factor in compensation for regional differences. These books can be very accurate, and they are an excellent way to evaluate prices given to you by suppliers and contractors.

My personal experience with cost-estimating manuals has shown that they usually predict estimates higher than most residential contractors give. I have figured jobs and then compared my figures with cost guides and found less than a $100 difference on big jobs; this is truly amazing. On the other hand, I have seen projections in guides for labor rates that were nearly double what most contractors were charging at the time in my area. I feel that the guides are useful, but I don't think you should rely on them as your sole source of information. If you are willing to spend the time to combine all these estimating methods, you can establish a very accurate estimate for the cost of your job.

> If you plan to use cost-estimating guides, use them in conjunction with the other ways of estimating your costs; don't depend on them exclusively.

4

Choosing Your
Materials

T he chore of selecting your materials can be perplexing. It can also be
a lot of fun. There are so many types of materials competing with each
other in the field of remodeling that even professionals have trouble
keeping up with what's available and what's best. For the average homeowner, trying to tell the difference between four different windows that
look the same but have huge differences in cost can be all but impossible.
The decision on whether to use wafer board or plywood for a subfloor
can cause hours of troubled thought. Asking yourself which type of faucet
will give the best service and appearance for the least amount of money
can drive you crazy. All these questions and more can arise when you are
trying to decide which materials to use. This chapter is going to show you
some simple ways to sort through the maze of products available.

The proper selection of materials can save a lot of money on the
overall cost of large remodeling jobs. There are some situations where
buying the best materials available will pay off and other occasions when
less expensive materials will get the job done and save you money

How can you, the homeowner, avoid costly mistakes? Research is the answer. If you ask enough questions and study enough product literature, you can avoid many pitfalls that people fall into. But you must know which questions to ask.

without sacrificing appearance or durability.

Knowing how to pick the proper materials is a skill remodelers often learn from trial-and-error experiences. While learning from experience is an effective method, it does get costly. And for homeowners it is unlikely that they will ever do enough remodeling to benefit from lessons learned the hard way.

SUBFLOORING

Subflooring is the flooring between the finished floor covering and the floor joists. The subflooring in most remodeling jobs does not need to be replaced, but there are times when it does. If the subflooring has been damaged by water or other causes, it may be necessary to replace it.

The three most common choices for subflooring material are plywood, wafer board, and a newer product that is much more resistant to water than either of the other options. This product goes by different brand names and comes in standard 4-x-8-foot sheets. If there will be only one layer of subflooring, the material should be of a tongue-and-groove (T & G) type. Some people install two layers of CDX plywood as subflooring, and others install one layer of tongue-and-groove plywood. If only one layer is being used, it is frequently ¾-inch T & G plywood. Most codes require this standard.

Wafer board is much less expensive than plywood—often half the cost. For this reason many professionals install a layer of wafer board and cover it with a thin layer of underlayment. By the time you add up the cost of the two layers and the time it takes to install them, one layer of T & G plywood is often less expensive. However, working with T & G materials in the confined spaces of a bathroom or kitchen can be troublesome.

It is one thing to build a new house using T & G throughout and quite another to fit it into the tight spaces available in most remodeling jobs.

The water-resistant material is more expensive than plywood or wafer board, but I use it. The added protection against water problems is worth it to me, and the cost is not staggering.

Figure 4.1   The materials chosen for this bathroom create a serene and peaceful environment. *Courtesy of Armstrong*

Talk to your contractor or supplier to see what brands of this type of material they recommend and have available.

Since bathroom remodeling does not require large quantities of subflooring, there will not be substantial money saved with any method or material. While installing two layers of subflooring may seem like twice as much work as installing one layer of T & G material, under remodeling conditions it is not. For most remodeling jobs it will be easier to install two layers of standard materials than one layer of T & G materials.

LUMBER

Lumber is available in different grades and at different prices. While the studs behind your walls won't be seen after the job is finished, their quality can affect the finished look of your project. Wood that twists and bends will warp the walls and can create defects in the finished wall surface. If you want to avoid problems with your lumber, insist on kiln-dried wood. For most applications a 2-grade lumber will be your best bet.

> If you have lumber delivered that will be kept outside, be sure to keep it dry. Letting rain get to your lumber can cause trouble down the road.

VINYL FLOORING

Vinyl flooring can range dramatically in price. The pricing spectrum may start at less than $15 per square yard and rise to a price in excess of $45 a square yard. An average grade of sheet vinyl flooring might be in the neighborhood of $18 per square yard.

Cheap vinyl can be difficult to install; it is not as flexible as better grades. Inexpensive vinyl is not likely to wear well, and it is

> Vinyl flooring should be warm when it is installed. If you are doing a job during cold weather, be sure to bring the flooring in before you plan to install it so that it can warm up.

prone to more cuts and tears. Vinyl flooring priced between $18 and $23 per square yard should be relatively easy to install, and it should hold up well.

BASE CABINETS

Cabinets are one of the most expensive elements of kitchen remodeling, and base cabinets account for a big part of this expense. There are many types of cabinets to consider, some of which include:

- Sink bases

- Drawer bases

- Bases with doors

- Bases with turntables

- Bases with pull-out trash receptacles

- The list goes on.

> When you buy cabinets, you will have to choose between dovetail joints and butt joints; dovetail joints should last longer.

If this isn't enough to confuse you, these base cabinets are made from a myriad of materials. Fortunately, bathrooms require far fewer cabinets. Many bathrooms contain a base cabinet for a vanity and possible a linen-type cabinet. All in all, bathrooms are less affected by cabinets than a kitchen would be.

Very few cabinets are made of solid wood; most contain some composite materials. If you want cabinets that are true solid-wood cabinets, be prepared to spend plenty of money. Most cabinets will have solid-wood fronts and plywood or particleboard interiors.

When you consider door options, you must decide if you want raised panels, flush doors, doors with finger pulls, or doors meant to accept hardware. This is a personal choice and shouldn't have a bearing on the durability of your job.

When checking out drawer bases, be sure to test how well the drawers slide in and out. Look for bases that have the drawers mounted on quality glides. This type of construction will provide years of trouble-free operation.

Investigate the structural integrity of the base cabinets. Good cabinets have supports in all corners, and the cabinets are firm. Cheap cabinets come in a you-assemble package with screws and little else. These units are the least desirable but also the least expensive.

Most people should avoid bargain cabinets that are sold unfinished. Finishing cabinets is not easy work, and it is easy to wind up with cabinets that don't match.

COUNTERTOPS

It is best not to purchase countertops until the base cabinets have been installed. This slows down the job, but it eliminates much of the risk of buying a countertop that doesn't fit. Coun-

Figure 4.2 Basic vanity base cabinets. *Courtesy of Wellborn Cabinet, Inc.*

**84" UTILITY
STORAGE
CABINET
2 Doors
1 Adj. full
depth shelf**

U1812
U1824

Figure 4.3   Utility cabinet. *Courtesy of Wellborn Cabinet, Inc.*

- Solid oak fronts.
- European-style self-closing hinges.
- Polished brass finish cabinet hardware.
- High-sheen honey oak finish.
- All exposed surfaces have woodgrain finish.
- Matching vanity, medicine
  cabinet, and light fixtures available.
- Fully assembled.
- Triple option hardware included to allow choice of
  brass, matching wood, or china.

Square Raised Panel – 7615183
18″ × 9¾″ × 29¾″ high
Triple option hardware included.

Square Flat Panel – 7616183
18″ × 9¾″ × 29¾″ high
Triple option hardware included.

Figure 4.4   Wall cabinets for a bathroom. *Courtesy of Universal Rundel*

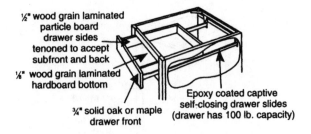

½" wood grain laminated
particle board
drawer sides
tenoned to accept
subfront and back

⅛" wood grain laminated
hardboard bottom

¾" solid oak or maple
drawer front

Epoxy coated captive
self-closing drawer slides
(drawer has 100 lb. capacity)

Figure 4.5   Vanity drawer specifications. *Courtesy of Wellborn Cabinet, Inc.*

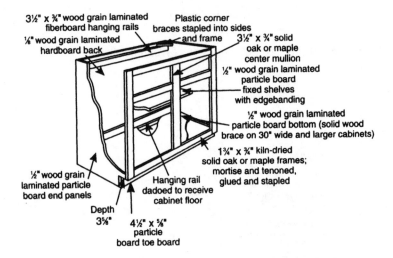

3½" x ¾" wood grain laminated
fiberboard hanging rails

⅛" wood grain laminated
hardboard back

Plastic corner
braces stapled into sides
and frame

3½" x ¾" solid
oak or maple
center mullion

½" wood grain laminated
particle board
fixed shelves
with edgebanding

½" wood grain laminated
particle board bottom (solid wood
brace on 30" wide and larger cabinets)

1¾" x ¾" kiln-dried
solid oak or maple frames;
mortise and tenoned,
glued and stapled

½" wood grain
laminated particle
board end panels

Hanging rail
dadoed to receive
cabinet floor

Depth
3⅝"

4½" x ⅝"
particle
board toe board

Figure 4.6   Base cabinet specifications. *Courtesy of Wellborn Cabinet, Inc.*

tertops cannot usually be returned for credit, so it is important
to get the right top on the first attempt.

If you are buying your cabinets and counters from a
custom cabinet company, you will not have to worry about
measurements. A representative will measure your kitchen and
decide on the proper sizes. However, if you are dealing with a
general supplier, you may have to measure for your own coun-

tertop. If this is the case, have the supplier explain to you exactly how to make the measurements for the type of top you are ordering.

Good, durable countertops are not terribly expensive, and they are available in a number of different colors and designs. Browsing through samples at your supplier will show you all the options. Expensive, specialty counters are not usually justified in average kitchens, but there are plenty of high-dollar tops available if you are so inclined.

WALL CABINETS

You can use the same basic rules applied to base cabinets for wall cabinets. Look for sturdy cabinets that offer a good appearance and adjustable shelves.

LAVATORIES

Lavatories fluctuate wildly in price. A simple wall-hung lavatory will cost less than $70, but a good pedestal lavatory can cost over $300. You can buy a plastic lavatory, a china lavatory, or an enameled cast-iron lavatory. China lavatories are the most common, and they hold up well.

Figure 4.7   A wall-hung lavatory with legs. *Courtesy of Eljer*

TOILETS

Can you imagine spending $1,500 for a toilet? There are many designer toilets available with big price tags, but a standard two-piece toilet will do fine, and it can be bought for less than $100. One-piece toilets look good, but they don't work any better than two-piece toilets. Once you go beyond the basics, you are paying for aesthetics, not functional durability.

BATHTUBS

Bathtubs can be made from plastic, fiberglass, enameled cast-iron, or enameled steel. Any of these tubs can provide years of dependable service, but the prices of the various types can differ considerably. A cast-iron bathtub may cost $600, while a steel tub that looks about the same can be bought for less than $175. With either of these types of tubs, a hard or sharp object can result in cracked enamel and a need for repairs. Also, both of these tubs will be much colder to sit down in than a plastic or fiberglass tub.

Fiberglass has become the most common type of bathtub. Fiberglass tubs are generally tough and well accepted. They are available as sectional units, which allow for a one-piece look from a modular unit. This can be a big boost to the remodeler who cannot get a standard one-piece tub-shower combination into a house. As for cost, they run between $350 and $450, but you will have the benefit of a tub and shower. They are, of course, available without the surrounding walls at a lower price.

> Cast-iron and steel bathtubs will be much colder to sit down in than plastic or fiberglass tubs.

SHOWERS

Most showers installed today are made of fiberglass. Like the modular tubs, showers are available in modular units. Some give

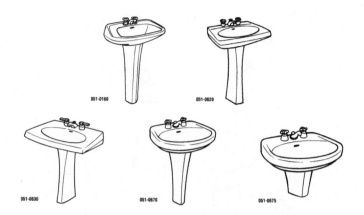

Figure 4.8 Pedestal lavatories. *Courtesy of Eljer*

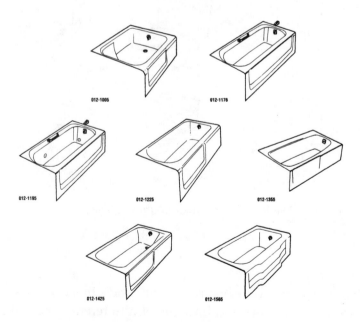

Figure 4.9 Cast-iron bathtubs. *Courtesy of Eljer*

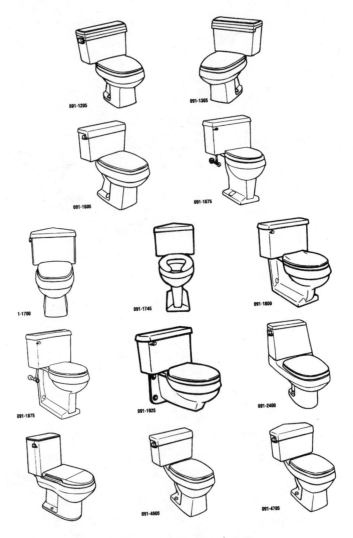

Figure 4.10   A variety of toilet types. *Courtesy of Eljer*

the appearance of a one-piece unit, and others use a shower base that is surrounded by walls of a different material. Typically, showers cost more the tub-shower combinations.

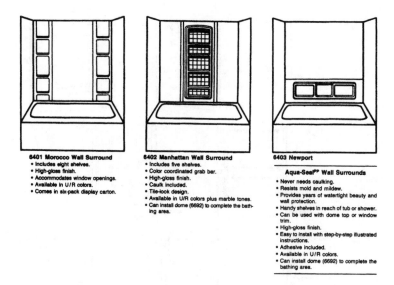

6401 Morocco Wall Surround
* Includes eight shelves.
* High-gloss finish.
* Accommodates window openings.
* Available in U/R colors.
* Comes in six-pack display carton.

6402 Manhattan Wall Surround
* Includes five shelves.
* Color coordinated grab bar.
* High-gloss finish.
* Caulk included.
* Tile-look design.
* Available in U/R colors plus marble tones.
* Can install dome (6692) to complete the bathing area.

6403 Newport

Aqua-Seal™ Wall Surrounds
* Never needs caulking.
* Resists mold and mildew.
* Provides years of watertight beauty and wall protection.
* Handy shelves in reach of tub or shower.
* Can be used with dome top or window trim.
* High-gloss finish.
* Easy to install with step-by-step illustrated instructions.
* Adhesive included.
* Available in U/R colors.
* Can install dome (6692) to complete the bathing area.

Figure 4.11   Basic tub surrounds. *Courtesy of Universal Rundel*

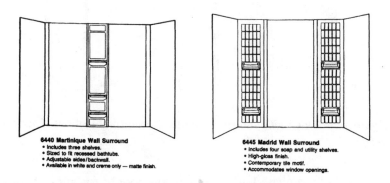

6440 Martinique Wall Surround
* Includes three shelves.
* Sized to fit recessed bathtubs.
* Adjustable sides/backwall.
* Available in white and creme only — matte finish.

6445 Madrid Wall Surround
* Includes four soap and utility shelves.
* High-gloss finish.
* Contemporary tile motif.
* Accommodates window openings.

Figure 4.12   Easy-to-install tub surrounds. *Courtesy of Universal Rundel*

LIGHT FIXTURES

Light fixtures are one type of material where you can spend a little or a lot and not really be able to see much difference. Recessed light fixtures can be bought for less than $40, but they are limited in the illumination they can provide.

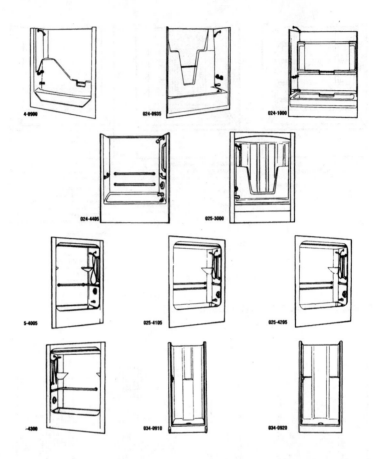

Figure 4.13   One-piece showers and tub-shower combinations. *Courtesy of Eljer*

Bar lights for the bathroom can be bought for about $25. If you want the same type of light in an oak finish, the cost will be around $30. This gives you an attractive bar light with three large designer bulbs for a very reasonable price.

Track lighting is very popular, and it can give plenty of illumination in different directions. Track lighting can be especially attractive and useful in a kitchen. Since track lighting is comprised of component parts, the price will vary with the

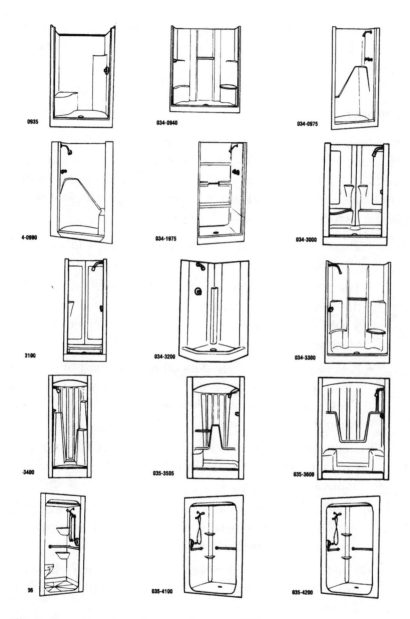

Figure 4.14 One-piece showers. *Courtesy of Eljer*

style you select and the number of housings installed on the track. All in all, track lighting is a very affordable way to obtain good lighting.

WINDOWS

Windows can be especially bothersome to evaluate. There are so many types and styles of windows to consider that your mind can stay confused for days. To eliminate some of this confusion, you can read product literature. Let's look at some of the factors you may want to consider:

- The energy-efficiency of windows is rated in terms of U-value.

- The lower the U-value is, the better and more efficient the window is. For example, a window with a U-value of 4 would be more energy-efficient than a window with a U-value of 5.

- Casement windows are generally considered one of the most energy-efficient types of windows you can buy. They crank open and allow a full flow of ventilation when the glass is open.

- Metal windows tend to sweat or condense. This moisture can cause damage to surrounding wood areas, and it is annoying visually.

- Wood windows provide good insulation and don't amplify sounds as much as metal windows do, but wood windows do require routine maintenance.

- Vinyl windows and vinyl-clad wood windows eliminate the need for painting. This maintenance-free feature is favored by many homeowners.

- Other options include gas-filled windows, low-E glass, glass that blocks UV rays, and so on.

Figure 4.15 This bathroom has exquisite fixtures and trim that give a feeling of luxury. *Courtesy of Armstrong*

There are so many possibilities that you must study brochures from numerous manufacturers to determine which features are of importance to you.

TRIM MATERIALS

The trim materials you install in your project are going to be one of the features that will affect the finished look of your new kitchen or bathroom. There are several choices available to you for trim. The two basic types of trim are finger-joint and clear. Finger-joint trim is less expensive than clear trim, but it is not suitable for staining. If you plan to paint your trim, finger-joint is fine. When you want to stain your trim wood, make sure it is clear trim or what is called stain-grade trim.

Colonial-style baseboards and casings are generally considered a good grade of standard trim. Clam-type baseboards are less expensive, but they are typically associated with lower-quality construction. You can use regular boards for trim. This is normally done when a rustic look is desired.

THE KEY TO CHOOSING THE RIGHT MATERIALS

The key to choosing the right materials is research. Educating yourself with product literature and books is an effective way to make sound purchasing decisions. If you don't understand what you are reading, ask professionals for their interpretations and opinions. If you allow adequate time for selecting your materials, you should be able to save money and still get good materials for your job.

5

Getting Your Best Price on Materials

S hopping for your best deal on materials will not be difficult once you know what you need and want. When you have a detailed list of material specifications, you can basically circulate the list among suppliers and wait for the best price. In theory, that's all there is to it. However, in real life, you will have to exert a little more energy to find the best bargains.

I had an occasion once to shop for materials with four different suppliers for the construction of a new house. I submitted identical plans and specifications to the suppliers and asked for detailed quotes. When all the quotes were in, there was a spread of about $4,000 from the lowest to the highest bid. This was many years ago when prices were much lower. Can you imagine that much difference among four suppliers for the same materials?

When I first saw the discrepancies I looked for errors in the bid sheets. I found none. After looking for mistakes, I began to compare the bids on an item-by-item basis. The picture started to come into focus as I studied the bid sheets. In some cases the materials I requested had been substituted, but for the most part, the difference was simply in the price of the materials. If this can happen to a professional contractor who buys materials on a daily basis, you can imagine how homeowners who may never visit a store again might be treated.

Aggressive shopping can save hundreds or even thousands of dollars on a kitchen or bathroom remodeling job. There comes a point where running from store to store is not cost-effective, but sometimes it pays to deal with more than one supplier. Many people feel that they will get the best deal by purchasing all their materials from the same vendor. While this should be true, it is not always the case. Let's look at some methods you can employ to get your best deal on materials.

Don't limit yourself to one supplier. Feel free to shop from many suppliers. It's okay to buy windows and doors from one supplier, lumber from another, and cabinets from yet another.

CIRCULATE YOUR PLANS AND SPECIFICATIONS

The first step in shopping prices is to circulate your plans and specifications. If you have a detailed take-off of the materials you want priced, that is all you need to distribute. Send the bid-request packages to several suppliers. If your job is a big one, it may pay to send bid packages to suppliers in other cities or states. It is amazing how much prices can differ from city to city and state to state.

If you do shop materials with long-distance suppliers, keep shipping costs in mind. They may outweigh any price advantage gained by buying at a distance. You may also want to consider the disadvantages long-distance suppliers offer in terms of product assistance and returns on damaged or improperly shipped items.

It is a good idea to call each supplier you will be sending a bid package to and request the name of an appropriate person to address the package to. Otherwise, your package may get shoved aside and neglected.

Once the bid packages have had time to arrive at the various suppliers, call and talk with your contact person. Confirm the delivery of your package and ask if there are any questions on the items you want priced. Inform the contact person not to make substitutions unless absolutely necessary, and if substitutions must be made, have the estimator note the changes in red ink. Insist that the bid be prepared in phases. For example, you might have the following phases for a bathroom job:

- Framing materials

- Wall coverings

- Plumbing fixtures

- Paint

- Cabinets

- Light fixtures

- And so on

> Inform the contact person at your supplier not to make substitutions unless absolutely necessary, and if substitutions must be made, have the estimator note the changes in red ink.

Having your bids broken down into distinct categories will make your overall evaluations easier.

GOING OVER THE QUOTES

Going over the quotes should be done when you have plenty of time and will not be disturbed. If you are in the middle of evaluating the prices and are frequently disturbed by a ringing telephone, you may overlook something of importance. The comparison process will work best if you have a large table to work at. This will allow you to lay all quotes next to each other for a quick, one-on-one comparison.

If all your quotes come back divided into the proper phases, consider yourself lucky. Suppliers don't like to go to the extra trouble to break down their bids, and they know you are more likely to spot high prices if you can compare prices side by side.

Most suppliers prefer to give a lump-sum figure at the end of a computer printout. When your quotes are not broken down properly, you will have to work a little harder to determine the real meat of the bid.

When you are reviewing quotes, do it in a quiet environment where you will not be disturbed. Avoid all distractions while you are focused on the details of quotes.

What you need to do is look at the price of a 2-x-4-inch stud from each supplier. Check each supplier's price per square yard for floor covering. There is little doubt that there will be ups and downs among these items. While it makes no sense to run all over town buying studs from one supplier, nails from another, and plywood from yet another, it does make sense to shop in phases.

Savvy remodelers deal with numerous suppliers and buy their materials in phases. This allows them to maximize their profits on materials. This same strategy can work for you.

Not all suppliers get the same discounts on all materials. One supplier may have a great price on cabinets and a terrible price on floor coverings. It will be no great inconvenience for you to buy flooring from one store and cabinets from another.

LOOKING AT THE BOTTOM LINE

Looking at the bottom line can be quite deceiving. If you look at four bids, it will be easy to see which vendor is offering the lowest overall price, but this doesn't mean it is the lowest price possible for the materials you need.

As you just learned, not all suppliers can sell the same products for the same price and make a profit even if they wanted to. There are usually some items that can be bought for less money at competitive suppliers.

By shopping the bids in phases, you will find the best prices available. You might have to deal with six different suppliers to save the most money, but you won't lose much time in doing so. There is nothing wrong in dealing with multiple suppliers.

WHAT'S INCLUDED?

What's included in each of the bids that you are comparing? Have all the suppliers included the sales tax on their bids? Generally, some suppliers will and others won't, and this can make a sizable difference in the bottom line. If the materials for your remodeling job will cost $10,000 and the sales tax in your area is 6 percent, the supplier who has included the cost of the tax will appear to be $600 higher than the suppliers who didn't include the tax. This situation is common and one you should look out for.

Will the suppliers deliver the materials for the prices quoted? This may not be a big deal if you have a truck and are only buying a few items, but if you are ordering for your whole job, delivery can be a problem and an expensive one at that. Most suppliers will deliver free of charge within a certain radius of their warehouse, but this is a question you should find the answer to before making a commitment.

Are the items you want regular in-stock items? If a supplier is quoting a job with materials that are not normally stocked, you may be faced with delays and extra hidden costs. When the supplier doesn't stock an item, you may have to pay for the shipping charges to have the item delivered to the supplier. It is unlikely that the supplier will reveal this expense in a competitive bid.

NEGOTIATING FROM PRELIMINARY PRICES

Once you have preliminary prices, you can begin the hardball negotiations. Most suppliers give professionals contractors at least a 10 percent discount on the prices of materials. This discount is rarely offered to homeowners. However, if the suppliers can sell to contractors for less, they can sell to you for less, but you will have to convince them to do so.

Let's assume that you have gone over all your bids and one supplier has offered you the best prices on everything except your floor covering and plumbing fixtures. You could buy from multiple suppliers to get the best price, or you could try negotiating with the one supplier that did well on pricing the other items.

> Push suppliers for a discount after you have gotten their best price. See if you can get 10 percent off the purchase price if you pay cash for the items. This can save you substantial money.

If you want to deal with just one supplier, take your other bids with you and go in for a personal visit with your sales representative. Explain how you would like to buy all of your materials from one store, but you can't justify the extra expense for the flooring and plumbing fixtures just for the sake of convenience. Ask the sales representative if anything can be done to lower the prices on the two overpriced items.

If the sales rep feels that the entire order is in jeopardy, there is a good chance that the sales manager will authorize a lower price on the flooring and plumbing fixtures. Remember, they probably have a 10 percent cushion of profit built into the entire order to work with. With a little persuasive negotiating, you can probably walk away with the best price and only have to deal with one vendor.

It may also be possible to negotiate for at least a portion of the discount normally offered to contractors. A lot of contractors finance their purchases for thirty days. Since you will most likely be paying in full upon delivery, use your payment terms as leverage. Impress the sales manager with the fact that there will be no risk in collecting a bad account and that the store is not really losing any money by giving you the same discount offered to professionals.

There are plenty of ways to drive a hard bargain if you are willing to spend the time and effort to do it. Suppliers don't want to see their competition get your business, and this gives you an edge. You don't really care where the material comes from, so you can shop until you get the deal you want.

> When dealing with suppliers, don't be afraid to ask for the price you are willing to pay. If the supplier knows that you are going to the local discount supplier for homeowners rather than pay the quoted price, you might see a discount applied to your quote quickly.

Once you have your materials secured for a good price, you may need to find and contract with some subcontractors to help you get the job done. If you will turn to the next chapter, you will see which subcontractors you might need.

6

Subcontractors

What do you know about subcontractors and their role in your project? Are you aware that the carpenter you hire to do your framing work may not have the skills or desire to do your trim work? Who hangs and sets the cabinets in the kitchen? Does a plumber install a dishwasher, or does an electrician do it? Do plumbing-and-heating companies deal with ductwork and furnaces? These questions are only a small sample of the types of questions commonly asked by homeowners.

If you are planning to act as your own general contractor, you must know what types of subcontractors will be needed for different phases of your job. Even small remodeling jobs can involve many different trades.

Once you get started on your job, you will probably become confused about the issue of subcontractors. Don't feel like you are the only one ever to be confused—many professional general contractors went through similar confusion when they got started in the business.

Today's working conditions have changed the way trades operate. It used to be that a carpenter could and would do almost any type of work that involved wood. Whether the job was making cabinets, building a barn, or hanging a door, just about any carpenter would handle the job. This is not the case today. Today's carpenters (and other tradespeople) are often highly specialized.

Since you will be searching for the best talent possible for your job, you must know what specialists to look for. Not being aware of the trends in specialization can create confusion and frustration. Imagine going through the phone book calling plumbers to help remodel your kitchen. What would your response be when you were turned down by the first five plumbers you called?

It's hard to believe a plumber would refuse to work on your remodeling job, but some will. If the plumbing company deals only in repair and service work, they will not be equipped to handle remodeling jobs. When the company does only commercial work, it will not be geared up for residential work. Plumbers who specialize in new construction will have no interest in a dirty remodeling job. To avoid this type of problem, let's take a look at each trade you may develop a need for.

TRASH CONTRACTORS

Most homeowners never think about hiring trash contractors until the rip-out debris begins to pile up. The pile of trash that accumulates during remodeling is surprising. There will be old lumber, flooring, fixtures, cardboard, and many other components of the job that must be disposed of. This is not a job that can be done reasonably using the trunk of your car. So who is going to get rid of the rubbish? This question has become harder to answer with the new rules at landfills and with the recycling efforts, but the solution is trash contractors.

Trash contractors may haul the unwanted material away in the bed of a truck or in a storage container. Most remodeling contractors rent a trash container that is placed on the job site. When the container is full, one phone call is all it takes to have it removed, emptied, and returned. There is little doubt that you will need to make some arrangements for the removal of debris.

SAMPLE SUBCONTRACTOR-SUPPLIED CONTRACT

Anytime Plumbing & Heating
126 OCEAN STREET
BEACHTOWN, ME 00390
(000) 123-4567

PROPOSAL CONTRACT

TO: Mr. and Mrs. Homeowner   Date: 8/17/04
ADDRESS: 52 Your Street Beachtown, ME 0039   PHONE: (000) 123-9876
JOB LOCATION: Same   JOB PHONE: Same   PLANS: Drawn by ACS, 4/14/04

ANYTIME PLUMBING & HEATING PROPOSES THE FOLLOWING:

Anytime Plumbing & Heating will supply and or coordinate all labor and material for the work referenced below:
PLUMBING
Supply and install a 3/4", type "L", copper water main from ten feet outside the foundation, to the location shown on the attached plans for the new addition.

Supply and install a 4", schedule 40, sewer main to the addition, from ten feet outside the foundation, to the location shown on the plans.

Supply and install schedule 40, steel gas pipe from the meter location, shown on the plans, to the furnace, in the attic, as shown on the plans.

Supply and install the following fixtures, as per plans, except as noted:
1 ABC Venus one piece, fiberglass, tub/shower unit, in white.
1 CF 007_222218 chrome tub/shower faucet.
1 ABC 900928 water closet combination, in white.
1 CBA 111 cultured marble, 30" vanity top, in white.
1 CF 005-95011 chrome lavatory faucet.
1 PKT 11122012 stainless steel, double bowl, kitchen sink.
1 CF 908001 chrome kitchen faucet.
1 DFG 62789 52 gallon, electric, 5 year warranty, water heater.
1 WTFC 20384 frost proof, anti-siphon silcock.
1 AWD 90576 3/4" backflow preventer.
1 FT66754W white, round front, water closet seat.
1 plastic washer box, with hose bibs.
Connect owner-supplied dishwasher.
All fixtures are subject to substitution with fixtures of similar quality, at Anytime Plumbing & Heating's discretion.

All water distribution pipe, after the water meter, will be Pex tubing, run under the slab. This is a change from the specifications and plans, in an attempt to reduce cost.

(Page 1 of 3) Initials_____

Figure 6.1 Example of a subcontractor-supplied agreement.

SAMPLE SUBCONTRACTOR-SUPPLIED CONTRACT (continued)

If water pipe is run as specified in the plans, the pipe will be, type "L" copper and there will be additional cost. Any additional cost will be added to the price listed in this proposal.

All waste and vent pipes will be schedule 40 PVC.

Anytime Plumbing & Heating will provide for trenching the inside of the foundation, for underground plumbing. If the trenching is complicated by rock, unusual depth, or other unknown factors, there will be additional charges. These charges will be for the extra work involved in the trenching.

All plumbing will be installed to comply with state and local codes. Plumbing installation may vary from the plumbing diagrams drawn on the plans.

Anytime Plumbing & Heating will provide roof flashings for all pipes penetrating the roof, but will not be responsible for their installation.

All required holes in the foundation will be provided by others.

All trenching, outside of the foundation, will be provided by others.

All gas piping, outside the structure, will be provided by others.

The price for this plumbing work will be, Four Thousand, Eighty Seven Dollars ($4,087.00).

HEATING

Anytime Plumbing & Heating will supply and install all duct work and registers, as per plans.

Anytime Plumbing & Heating will supply and install a BTDY-P5HSD12N07501 gas fired, forced hot air furnace. The installation will be, as per plans. The homeowner will provide adequate access for this installation.

Venting for the clothes dryer and exhaust fan is not included in this price. The venting will be done at additional charge, if requested.

No air conditioning work is included.

The price for the heating work will be Three Thousand, Eight Hundred Dollars ($3,800.00).

Any alterations to this contract will only be valid, if in writing and signed by all parties. Verbal arrangements will not be binding.

PAYMENT WILL BE AS FOLLOWS:

Contract Price of: Seven Thousand, Eight Hundred Eighty-Seven Dollars ($7,887.00), to be paid; one third ($2,629.00) at the signing of the contract. One third ($2,627.00) when the plumbing and heating is roughed-in. One third ($2,629.00) when work is completed. All payments shall be made within five business days of the invoice date.

(Page 2 of 3) Initials_____

Figure 6.1 (continued)   Example of a subcontractor-supplied agreement.

SAMPLE SUBCONTRACTOR-SUPPLIED CONTRACT (continued)

If payment is not made according to the terms above, Anytime Plumbing & Heating will have the following rights and remedies. Anytime Plumbing & Heating may charge a monthly service charge of one percent (1%), twelve percent (12%) per year, from the first day default is made. Anytime Plumbing & Heating may lien the property where the work has been done. Anytime Plumbing & Heating may use all legal methods in the collection of monies owed to Anytime Plumbing & Heating. Anytime Plumbing & Heating may seek compensation, at the rate of $50.00 per hour, for their employees attempting to collect unpaid monies. Anytime Plumbing & Heating may seek payment for legal fees and other costs of collection, to the full extent that law allows.

If Anytime Plumbing & Heating is requested to send men or material to a job by their customer or their customer's representative, the following policy shall apply. If a job is not ready for the service or material requested, and the delay is not due to Anytime Plumbing & Heating's actions, Anytime Plumbing & Heating may charge the customer for their efforts in complying with the customer's request. This charge will be at a rate of $50.00 per hour, per man, including travel time.

If you have any questions or don't understand this proposal, seek professional advice. Upon acceptance this becomes a binding contract between both parties.

Respectfully submitted,

H. P. Contractor
Owner

PROPOSAL EXPIRES IN 30 DAYS, IF NOT ACCEPTED BY ALL PARTIES

ACCEPTANCE
We the undersigned, do hereby agree to and accept all the terms and conditions of this proposal. We fully understand the terms and conditions and hereby consent to enter into this contract.

Anytime Plumbing & Heating   Customer #1
by_____   _____
Title_____   Date_____

Date_____   Customer #2
                                _____
                                Date_____

(Page 3 of 3)

Figure 6.1 (*continued*)   Example of a subcontractor-supplied agreement.

ROUGH CARPENTERS

Rough carpenters usually do framing work. If you were building a new house, you might have a crew of rough carpenters erect the shell. In the case of remodeling, you might find one carpenter to do all the work involved, or you may have to look for different types of carpenters. If you are not expanding the size of your bathroom, you will not have much need for rough carpenters.

Some carpenters are not what they seem. It is not unusual for people to lose their jobs. When they do, some of them go out and buy a hammer so that they can call themselves carpenters. You don't want a rookie doing your remodeling. Make sure that all your contractors are experienced, licensed, and insured.

TRIM CARPENTERS

Trim carpenters are the people who hang doors, install baseboard trim and window casing, and so forth. These carpenters are good with detail work. While many of them work slowly, they often work to perfection. Some trim carpenters will install cabinets, but not all of them will. You will almost certainly need a trim carpenter.

CABINET INSTALLERS

Cabinet installers may do nothing but install cabinets. These people frequently work for companies that sell cabinets. If you were to go searching for a subcontractor to install your cabinets, you should start by asking the supplier of your cabinets for recommendations.

INSULATION CONTRACTORS

Insulation contractors are not always needed on bathroom remodeling jobs. However, if the job involves building an addition onto the home or something along those lines, an insulation contractor may be wanted. Most insulation contractors have separate crews for different types of work. If you call an insulation company, your rep should be able to organize a crew to do the work you request.

DRYWALL HANGERS

On small jobs drywall hangers often take care of the finishing work. However, on large jobs it is not uncommon to hire one subcontractor to hang drywall and another to finish it for paint.

Custom cabinets can take many months to build. Getting cabinets in two months is fast for custom jobs. I talked with an employee of a custom-cabinet company a few weeks ago who told me that his company had a waiting list that was nearly a year long. Plan your cabinet delivery well in advance.

DRYWALL FINISHERS

Drywall finishers specialize in finishing drywall. While it would be unusual to use separate contractors to hang and finish drywall on a small job, it may be to your advantage. People who spend every day

Most homeowners can install their own insulation, but many people are allergic to insulation products. If you decide to install your own insulation, protect yourself with clothing and gloves that will cover your skin. It is also wise to wear a dust mask when working with insulation.

finishing walls and ceilings are naturally going to be better at it than people who spend half their time hanging drywall.

WALLPAPER HANGERS

Homeowners can hang wallpaper, but wallpaper hangers specialize in this field of expertise. Using professional hangers should result in a more uniform and appealing job. If you are

Drywall work is a critical part of a successful remodeling job. Make sure that your drywall contractor is reputable and check the work very closely before you pay for it.

not experienced in working with wallpaper, it is easy to mismatch patterns and seams.

PAINTERS

Who needs painters? Anybody can paint walls and ceilings, can't they? Almost anyone can paint walls and ceilings, but professional painters tend to do a much better job than the average person would. There are many tricks of the trade when it comes to painting, and your job will benefit from the expertise of experienced painters.

I can't begin to count the number of homeowners who have told me that they can do their own painting. It seems that nearly everyone assumes that anyone can paint. While it is true that almost anyone can apply paint to a wall or a ceiling, this does not mean that these people are real painters. Painting is more complicated than it looks. This is something that you can probably do, but don't underestimate the time and skill required for a professional painting job.

ELECTRICIANS

Electricians all work with electricity, but they don't all do it in the same ways. Some electricians do only commercial work, and others do only repair work. Neither of these types of electricians will be the best choice for a remodeling job. What you need is an electrician who does residential work and preferably remodeling work. There are substantial differences between wiring a new house and wiring a remodeling job. You should spend the extra effort to find an electrician with remodeling experience.

HEATING MECHANICS

You may not have any need for heating mechanics on your job. If you are not adding space or relocating existing heating units, you won't have a need for these tradespeople. If you do have heating needs, look for mechanics who are experienced in residential remodeling. Heating mechanics are similar to electricians and plumbers in their divisions of specialization.

SUBCONTRACTOR SCHEDULE

Type of Service	Vendor Name	Phone Number	Date Scheduled

Notes/Changes:

Figure 6.2 Example of a contractor schedule form.

PLUMBERS

Plumbers specialize in all types of work. You can find plumbers who specialize in any of the following types of jobs:

- Sewer cleaning
- Well systems
- Water conditioning systems
- New construction
- Commercial work
- Residential work
- Repair work
- Remodeling
- Water-service and sewer installations

Find a plumber who knows residential remodeling inside and out.

FLOORING INSTALLERS

Flooring installers are often available through the stores that sell floor coverings. The flooring installer you hired last year to replace the carpeting in your family room may not install vinyl flooring. While most flooring installers will work with any type of carpet or vinyl, it pays to make sure you are getting someone with experience in the type of work you require.

TILE CONTRACTORS

If you want tile installed, you may need to contact tile contractors. These individuals can be found in phone directories and through the store supplying your tile.

SIDING CONTRACTORS

If you are adding space onto your home or installing new windows, you may have a need for siding contractors. Many carpenters will work with siding, but there are companies that specialize in siding work.

ROOFERS

It is unlikely that you will need roofers for an average bath remodel, but if you are installing skylights, bay windows, or adding space, you might. Most carpenters are willing to do minor roofing jobs, but if the work is extensive, it may be less expensive to deal directly with a roofing contractor.

Now that you have a good idea of the types of subcontractors you may need, let's move on to the next chapter and see how to select the best subs for the job.

Roofing work is not included in the insurance polices carried by some carpenters. If you want a carpenter to do roofing work, make sure that the contractor has the proper insurance for the job.

7

Selecting Contractors and Subcontractors

I t is very likely that you will need at least a few subcontractors when remodeling your bathroom. Selecting your subs is a job in itself. Choosing the wrong people will turn your dream job into a nightmare. If you will be hiring a general contractor to handle the entire job for you, the burden of subcontractors will be lifted off your shoulders. However, you will still have to exercise prudence in picking a general contractor. Any way you cut it, you are going to have to deal with contractors unless you do all aspects of the job yourself.

Finding and selecting the right subcontractors are not easy tasks. I say this from experience. At one point, I was building up to sixty homes per year. That's a lot of houses and it requires a lot of help. I had some tradespeople on payroll, but subcontractors provided most of the trade labor. As I recall, I had over 120 subcontractors and vendors in any given year.

I started in the construction industry as a plumber, and my first business was a plumbing business. During that time I was a subcontractor. As my experience and business grew, I moved into remodeling and building;

it was then that I became a general contractor. Having been a subcontractor and having worked around other subcontractors, I had more experience on the subject of subcontractors than most new general contractors. Even so, finding, choosing, and keeping the best subs was a chore, and it still is.

I have given you this brief background to prove two points: I have extensive experience in working with subcontractors, and if a professional has to continue to work at finding and selecting good subs, a homeowner should not take the task lightly.

WHERE SHOULD YOU LOOK FOR SUBCONTRACTORS?

Where should you look for subcontractors? The advertising pages of your local phone book are a logical place to start. Contractors listed in the phone book have been established in business for at least a little while, and they are easy to locate.

> Don't assume that all contractors who have ads in the phone book are reputable. It only takes money to buy ads. The skills that your job will require include much more than money. But a contractor who is listed in the phone book is a good place to start.

Newspapers

Local newspapers are another good place to look for subcontractors. The rates for advertising in phone books are steep, and some good contractors have nothing more than a line listing in the phone directory. These contractors prefer to spend their advertising budget selectively, and the newspaper is one of the places they may use to announce their services to the public.

Friends

Ask your friends if they know of any reputable contractors. It is hard to beat trusted word-of-mouth referrals when searching for good contractors.

Look for Signs

When you are riding around your neighborhood, look for signs that indicate the presence of remodelers working in your area. Many contractors display signs with their company names and phone numbers when doing a job. Finding contractors this way shows you that they are working, and you can probably gain permission to inspect the work they are involved with for others.

> When checking the references of contractors, ask to be shown some of the jobs that they have completed, but also ask to see jobs in progress. The jobs in progress will be easier to gain access to, and it will give you a better view of the work as it is being done at the present time. This is an advantage.

WHAT QUALITIES SHOULD YOU LOOK FOR IN SUBCONTRACTORS?

What qualities should you look for in subcontractors? In some cases the qualities will vary with the different types of contractors, but there are some common attributes to seek.

Licensing

Any contractor you hire should be licensed to conduct business. Many trades, such as plumbing, heating, and electrical contracting, require special licensing. For example, a plumbing contractor should hold a master plumber's license. Having a journeyman plumbing license is not sufficient for most plumbing contracting.

Check with your local code and licensing authorities to determine which licenses are required for various types of work, and don't associate yourself with contractors who are not properly licensed.

CONTRACTOR QUESTIONNAIRE

PLEASE ANSWER ALL THE FOLLOWING QUESTIONS, AND EXPLAIN ANY "NO" ANSWERS.

Company name _____

Physical company address _____

Company mailing address _____

Company phone number _____

After hours phone number _____

Company President/Owner _____

President/Owner address _____

President/Owner phone number _____

How long has company been in business? _____

Name of insurance company _____

Insurance company phone number _____

Does company have liability insurance? _____

Amount of liability insurance coverage _____

Does company have Workman's Comp. insurance? _____

Type of work company is licensed to do _____

List Business or other license numbers _____

Where are licenses held? _____

If applicable, are all workman licensed? _____

Are there any lawsuits pending against the company? _____

Has the company ever been sued? _____

Does the company use subcontractors? _____

Is the company bonded? _____

Who is the company bonded with? _____

Has the company ever had complaints filed against it? _____

Are there any judgments against the company? _____

Please list 3 references of work similar to ours:

#1 _____

#2 _____

#3 _____

Please list 3 credit references:

#1 _____

#2 _____

#3 _____

Please list 3 trade references:

#1 _____

#2 _____

#3 _____

Please note any information you feel will influence our decision:

ALL OF THE ABOVE INFORMATION IS TRUE AND ACCURATE AS OF THIS DATE.

DATE:_____   COMPANY NAME: _____

BY:_____   TITLE: _____

Figure 7.1   Example of a questionnaire for a contractor.

CONTRACTOR RATING SHEET

Job name: _____ Date: _____

Category	Contractor 1	Contractor 2	Contractor 3
Contractor name			
Returns calls			
Licensed			
Insured			
Bonded			
References			
Price			
Experience			
Years in business			
Work quality			
Availability			
Deposit required			
Detailed quote			
Personality			
Punctual			
Gut reaction			

Notes: _____

_____

_____

_____

_____

_____

_____

Figure 7.2 Example of a contractor rating form.

Insurance

Check to see that any contractors being considered for your job are properly insured. Insurance for contractors is expensive, and many contractors don't have insurance. Hiring an uninsured contractor is very risky business.

When you verify a contractor's insurance, verify it with the agency issuing the insurance policy; don't take the contractor's word for it. If you are reviewing a certificate of insurance on a contractor, examine the dates carefully. The policy may have expired.

Experience

Experience in remodeling is vital to the successful completion of your job. It is easy for contractors to say they have years of experience with remodeling work; make them prove it. Contractors who work with new construction are not always qualified remodelers. The differences between remodeling and new work are significant.

References

Since you will be requiring contractors to prove their experience in remodeling, check the references they provide you with. Insist on at least five references. Shady contractors will be expecting you to ask for three references, and they may have three friends or relatives prepared to pose as references. By asking for five references you may be able to catch crooked contractors off guard. Ideally, you should visit jobs the contractors are doing to see that the references you are given are real.

You can hire a very experienced contractor and still have the wrong contractor for your job. Make sure that the contractor's experience is in the field related to your type of work. In short, don't hire a commercial heating company to relocate your residential ductwork. The commercial crew may be able to do the job without hesitation, but you should be better off with a company experienced in residential remodeling.

Business Stability

Business stability in remodeling and contracting can be difficult to maintain. Swings in the economy can turn a company that was thriving the year before into a bankrupt business. Some businesses that present a strong public image can be on the verge of collapse. You trust your money and your house to these contractors, and you are entitled to know that your trust is not misplaced.

> Dig deep and find out as much as you can about the contractors and the stability of their businesses.

HOW CAN YOU VERIFY INFORMATION GIVEN TO YOU BY SUBS?

How can you verify information given to you by subs? Insurance information can be verified by calling the insurance agency providing coverage to the subcontractor. References can be called, but whenever possible they should be visited. Licensing information can be verified with the licensing agencies in your area. Experience is difficult to verify, but checking references will provide some security.

If the contractor will give permission for you to talk with material suppliers, the suppliers can tell you much about the stability of the contractor's business. Requesting permission to talk with the contractor's bank is another way to check for business stability.

Many contractors belong to professional organizations that may be willing to provide some background information. Credit reports on the contractors would be of help, but few contractors are going to provide homeowners with the same verifications they will offer a professional general contractor.

Getting answers to all your questions will not be easy. Many subcontractors will consider your requests unusual and more trouble than your business is worth. Since most homeowners ask very few questions of contractors, the few that do face an uphill battle. While you can't expect to be completely safe working with subs, you can hedge the odds by learning as much about the contractors as possible.

CHOOSE BACKUP SUBCONTRACTORS RIGHT FROM THE START

When you begin selecting subs, choose backup subcontractors right from the start. Invariably, there are times when the subcontractors you want to do the job will not work out. Sometimes they will be injured and unable to perform the work. There will be occasions when the contractors are behind schedule and cannot get to your job when they are supposed to. Some contractors will seem to just disappear, and you won't be able to find them to do the work. It is better that this happens before work is started than after. There are dozens of reasons for selecting backup contractors.

> Wise professionals always have at least three subcontractors for every phase of work. Having three electricians chosen in advance will allow more freedom and less chance for a catastrophe than putting all your trust in one subcontractor.

When you are sorting through potential contractors, give them ratings. For example, when considering drywall contractors, establish a first, second, and third choice for your job. This type of advance planning may save you much trouble once the job is started.

BEWARE OF SALES HYPE

Beware of sales hype. Some contractors are better at selling their services than providing them. As a consumer, you must screen all contractors carefully and avoid camouflaged salespeople. The best salespeople will not appear to be selling you anything. True professionals can make you buy products and services you don't need and don't want. Learning this from experience can be very costly.

CONTRACTOR COMPARISON SHEET

Category	Contractor 1	Contractor 2	Contractor 3
CONTRACTOR NAME			
RETURNS CALLS			
LICENSED			
INSURED			
BONDED			
REFERENCES			
PRICE			
EXPERIENCE			
YEARS IN BUSINESS			
WORK QUALITY			
AVAILABILITY			
DEPOSIT REQUIRED			
DETAILED QUOTE			
PERSONALITY			
PUNCTUAL			
GUT REACTION			

Notes: _____

_____

_____

_____

_____

_____

Figure 7.3 Example of a contractor comparison form.

If you are told you should remodel your bathroom at the same time that you are having your kitchen remodeled, think the proposition over. There is a good chance that you could save money by having both jobs done simultaneously, but if your bathroom doesn't need to be remodeled, you are wasting your money.

You can protect yourself from impulse buying by refusing to make an on-the-spot decision. When you are told that if you don't act immediately the price will go up, look for another contractor.

SATISFACTION

To achieve satisfaction from your completed job, you may need dependable, reputable, experienced subcontractors. Spend enough time in the selection process to ensure your satisfaction. The wrong contractors can turn the best job into a horror show, but the right contractors can make difficult work look easy.

8

Dealing with Contractors

How well you deal with contractors can have a strong influence on the cost of your job. The prices contractors quote for similar work can span a broad spectrum. The same could be said for any subcontractors you might need. Even if the difference in hourly rates is only $10, the cost over the course of the job can amount to quite a bit.

Let's look at a quick example of how the hourly rates used by two plumbers in estimating a bathroom-remodeling job might affect your costs. One plumber is pricing labor at $52 an hour, and the other is figuring $62 an hour. Both plumbers are figuring the same amount of time for the work required.

When the plumbers estimate the time they will devote to removing existing fixtures, they figure five hours plus travel and miscellaneous time. To round it off, they both consider the rip-out to be worth eight hours of their time.

When figuring the labor for setting new fixtures, both plumbers arrive at an estimate of another eight hours. Allowing for inspections, handling material acquisitions, and other administrative duties will be billed as an additional eight hours.

Using broad-brush estimating techniques, both plumbers have arrived at estimates of 24 hours for their involvement in your remodeling job. The highest estimate is $1,488 and the lowest estimate is $1,248. There is about a 17 percent difference in the costs, and this is just for the plumbing phase of the job.

Assume that the total cost for all labor in your bathroom-remodeling job will cost $6,000. If you could shave 17 percent off that labor figure, you would save $1,020. Is it possible to save this much money and still get a good job? Yes, it is possible, but it will require effort on your part.

Finding the best deals on labor takes patience and persistence. Contractors are not going to give their services away, but they may be willing to discount them heavily if you know the right buttons to push. There are ways to convince contractors to give you a lower price for the privilege of getting your job. Why would contractors do this for you? That's what you are about to find out.

> Think twice before you agree to let a contractor use your job as fill-in work. This is rarely a good idea for kitchen remodeling or when a home has only one bathroom. Why? You could be without your facilities for much longer than you wish to be.

FILL-IN WORK

All contractors love to have fill-in work. This is work they can do when circumstances prohibit them from doing regularly scheduled work. If it is a rainy day and a carpenter cannot work outside, having a fill-in job that allows inside work is a welcome pleasure. If a delivery is mixed up on one job and brings a crew to a halt, the contractor will not lose as much money if the crew can be sent to work on a fill-in job.

Since fill-in jobs are valuable to contractors, you can often negotiate for a lower price if you are willing to allow your work to be a fill-in job. This works well with some types of home improvements, but it usually is not a good idea with kitchen remodeling. If you have more than one bathroom, you might be able to consider having your additional bathroom remodeled on a fill-in basis, but be prepared to do without the use of that bathroom for an extended period.

Due to the nature of fill-in jobs, they don't get finished quickly. If you are willing to allow your work to drag out for weeks or perhaps months, you are in a good position to bargain for a lower price. However, if you need the work completely in a timely fashion, don't consider a fill-in job.

REFERENCE JOBS

Reference jobs are not always easy for contractors to come by. What is a reference job? A reference job is a job that a contractor can give as a reference to future customers. If appointments can be scheduled with the homeowners of completed jobs for showing the work to prospective customers, contractors can close more sales. Any progressive contractor recognizes the value of having jobs that can be shown to potential customers.

If you structure a deal with contractors to use your house as an example of their work, you win two ways. First, you win by getting a lower price for the services you receive. Secondly, you win because, since the job will be used to show off the contractors' talents, you will get the benefit of the best job the contractors can do.

Allowing your job to be used as a showplace is a very effective way to negotiate for better pricing and sometimes for better materials. Remember, the contractors will be showing the job to future customers, and the business owners

Just as you should arrange to see samples of a contractor's work, so should other customers. If you are willing to allow contractors to use your job as a reference job, you should receive something for the inconvenience of being a good reference for the contractor. This something can often be a lower price for the work being done.

New companies are frequently starved for work and references. Dealing with the right new companies can save you considerable money. If you happen to deal with the wrong new companies, the pain will outweigh the advantages; choose your contractors carefully.

will want their new customers to be favorably impressed. However, if you do agree to such an arrangement, establish the terms in writing. Dictate how much notice you will be given prior to showings, how long your job will be used as a reference, and how many showings will be allowed on a weekly basis.

NEW COMPANIES

New companies are frequently starved for work and references. Dealing with the right new companies can save you considerable money. If you happen to deal with the wrong new companies, the pain will outweigh the advantages; choose your contractors carefully.

There are risks to dealing with fledgling companies, but there are also rewards. If the contractors are honest and experienced, you can get better service than you might from a large, established company, and the price will almost always be less.

When contractors first go into business for themselves, they need work badly. Many of these new business owners are not new to their business, only to owning it. The contractors could have twenty years of experience doing their job but have only been in business for a month.

Should you decide to gamble on using a new company? Do your homework. Make sure that the business is insured and licensed. While there may not be much to investigate, dig up everything you can on the company and its owner.

NEGOTIATE

Don't be afraid to negotiate with contractors. When you go to buy a new car, do you automatically agree to pay the full price on the window sticker? If you haggle over the price of a car, why shouldn't you negotiate for a better price from contractors? In many cases the full price of a major remodeling job is more than the cost of an average new car, so certainly there is enough money involved to be worth negotiating. Many homeowners never question the prices given by contractors. They either look elsewhere for a lower price or accept the price they are given.

I have been involved in construction for about thirty years. Many of these years have been as a contractor, and I have rarely had homeowners dicker with me on prices. As a subcontractor I have had plenty of general contractors bargain for better prices, but few homeowners ever do. This is a situation I have never really understood.

Very few contractors show all their cards on the initial bid. They bid jobs expecting to negotiate. When homeowners accept the bid as is, the contractors pocket some extra profits. If homeowners look at the prices and continue to search for other, lower-priced contractors, the contractors with the padded bids lose out.

> If you have found a contractor who fits the perfect profile, don't let a high bid alienate you. Good contractors are hard to find, and when you do find one, you should make an effort to work out a viable deal. Contractors are not so different from other business owners that they are unwilling to consider offers and negotiations.

While many good contractors don't inflate their prices enormously, most do build in a buffer of at least 5 percent. It is not unusual for contractors to inflate their prices by 10 percent. Knowing this, you should try to squeeze that price cushion out of the contractors. Saving 5 percent on a $20,000 kitchen job means saving $1,000. While $1,000 is not enough to retire on, it is enough to pay for a new microwave and dishwasher.

REMODEL WITH THE SEASONS

When you remodel with the seasons of the year, you can save money. There are certain times of the year when contractors traditionally have less work than they would like. If you can arrange to have your job done during the off-season, many contractors will reward you with a lower price.

What is the best time of the year to have your remodeling done? There are several good times of the year for you to take advantage of special prices and incentives. Let's start with the first of the year and look at each month to establish your best buying times.

Following are the months of the year and the pros and cons of each for getting your best deal:

- January: January is an excellent time to schedule your remodeling work. Many homeowners are still recovering from the holiday season and adjusting to the new tax year. In most regions the weather during January is not conducive to construction and happy attitudes. This makes the first month of the year a good time to save some money on your remodeling costs.

- February: February is often cold and dreary—an excellent time to offer contractors inside work. However, February is also the month when many contractors are getting geared up for the spring rush, so they may not be quite as flexible as they would be in January.

- March: March is a month when the odds favor contractors. People who have been depressed through the winter see a glimmer of hope for warm, sunny days, and they are more likely to have a positive outlook. This change in attitude often results in buying things: home improvements, houses, cars, and so on. Avoid the month of March.

- April: April is another month to avoid. By April, people are sure summer is getting closer, and they are ready to go on spending sprees. This is not a good time to look for bargains in home improvements.

- May: May does not offer any better opportunities than April. The spring months are the worst choices for home-improvement values.

- June: June is not appreciably better than April or May; it is another month to avoid.

- July: July can offer some windows of opportunity for bargain shoppers. Many people vacation in July, and this results in less work for contractors. The heat of July can also make contractors look for inside work, where air conditioning will temper the heat. While July is not the hottest month for home-improvement values, it is better than the months of spring.

- August: August is a good time to have inside work done. Oppressive heat in many parts of the country makes contractors dream of working in air-conditioned spaces. Another advantage to August is that many parents are getting their children ready to go back to school or college. The preoccupation with the costs and duties of getting the kids back in school prevents these people from pursuing remodeling work. Their lack of interest in remodeling during August will work to your advantage.

- September: September is no good for discount remodeling prices. This is a month when many people decide to remodel before winter, so avoid shopping in September.

- October: October is worse than September for bargain hunting. People are rushing about to get their homes in order before the holidays, and contractors have no shortage of work during these times.

- November: By the middle of November, the remodeling market becomes a buyer's market. Most people don't want contractors in their homes during Thanksgiving, and this opens the door of opportunity to lower prices for you.

- December: December is an excellent month for low remodeling prices. Most people don't contemplate remodeling between Thanksgiving and Christmas. This gives you ample opportunity to cash in with some hefty savings. It also doesn't hurt that contractors often want extra money for Christmas.

A creative homeowner can find numerous levers to manipulate contractors into lower prices. If you put your mind to it, you should be able to save substantial money on your job by strategic shopping and negotiating.

9

Code Considerations

Permits and code compliance are considerations that all homeowners involved with remodeling projects should be aware of. If you will be doing your own remodeling work, these aspects of the job are of the utmost importance. When you hire a general contractor to do the job for you, you shouldn't have to worry about code compliance and permits. But if you are smart, you will take the time to learn at least in general terms what should and shouldn't be done.

Unfortunately, not all contractors play by the rules. Some contractors avoid getting permits for their jobs to keep the price of the work lower. A few contractors are unable to get permits because they are not licensed and therefore never inform their customers of a need for permits and inspections. For whatever the reason, it is not a good idea to perform work that requires a permit and inspection without the proper authorization.

Any work done that does not comply with code requirements may have to be torn out and done over again. If you were unlucky enough to hook up with a bad contractor, you might wind up paying for the contractor's mistakes, and this can get expensive very quickly.

Almost every town, city, and county has a code-enforcement office and official code inspectors. The inspectors are employed to protect the citizens of the jurisdiction, including you. While arranging for code inspections may not be something you want to do, it may be required by law in your area and the inspection may protect you from all sorts of dangers.

Require all contractors who are responsible for the acquisition of permits and inspections to furnish you with copies of the inspection approvals. Don't pay completion fees to contractors without proof that their work has been inspected and approved, as required by your local code-enforcement office.

If you will be doing work that requires a permit yourself, you can apply for one as a homeowner. Owners who reside in their property can be issued permits that would normally require professional licensing.

Failure to comply with code regulations can result in fines and sometimes imprisonment.

DOES ALL WORK HAVE TO BE INSPECTED?

Does all work have to be inspected? No, not all work requires a permit or an inspection, but much of the work involved with bathroom remodeling does require permits and inspections.

When permits are required, they must be obtained before any work regulated under the permit is started. The permits are normally required to be posted on the job in a conspicuous place where they can be seen from the street.

WHAT'S INVOLVED IN GETTING A PERMIT?

What's involved with getting a permit? Getting a permit is not difficult, but there are guidelines that must be followed. The procedure for obtaining a permit can vary from jurisdiction to

Figure 9.1   This exotic look can be achieved with the expertise of a good contractor. *Courtesy of Armstrong*

jurisdiction. Check with your local code authorities for exact details on what the requirements are in your area.

Typically, a set of plans and specifications must be submitted to the code-enforcement office, along with a completed permit application (available from the code office) to receive a permit. Payment for the permit is also required.

Some types of work may not require plans and specifications to be submitted. Under these conditions, the only paperwork involved will be the permit application. There will be a fee for the permit if it is issued. The amounts of fees vary from location to location, but the costs for all the permits needed for a major remodeling job can run into hundreds of dollars. Most single permits will be less than $75, and many will be less than $50.

If you apply for your own permits, be advised that normally permits are only issued to homeowners who will be

Do not obtain permits in your name and then allow unlicensed people to do the work for you. This could land you in significant trouble. If you are thinking of skirting the issue of an unlicensed contractor by getting the permit yourself, forget it. The risk is not worth it.

performing the work themselves and to licensed contractors. If you apply for the permit in your name and hire an unlicensed person to do the work for you, there could be trouble if you are caught.

Once the permit has been applied for, a code officer will review the application. If all the paperwork is in order, the permit fee will be paid (by you or your contractor) and the permit will be issued. The permit should be posted at your home in such a way that an inspector can see it from the street.

WHAT TYPES OF WORK REQUIRE PERMITS?

What types of work require permits? Most small jobs that are mostly repairs or maintenance don't normally require permits and inspections, but larger jobs do. Let's look at where permits are likely to be needed for your job:

- Plumbing: Plumbing work almost always requires a permit and inspection. Unless all you are doing is very minor remodeling or repair work, plan on getting a plumbing permit. Normally, if you are relocating fixtures, adding fixtures, or doing extensive work with plumbing pipes, a permit and inspections will be required.

- Electrical: Electrical work is similar to plumbing in terms of permits and inspections. If you are adding a circuit, relocating wiring, or work of this nature, a permit will probably be required.

- HVAC: Heating and air-conditioning work is also normally done with the use of a permit. Very minor changes will not require a permit or inspection, but if you are expanding your HVAC system or making significant changes in outlet locations, you may need a permit.

Figure 9.2   Permits are necessary when remodeling to achieve a look such as this beautiful bathroom. *Courtesy of Armstrong*

- Building: Building permits are normally required whenever a new structure is built, but they may not be required for cosmetic remodeling. However, if you will be making structural changes to your home, a permit is likely to be required.

To be safe, you should inquire at your local code-enforcement office for information on what is required of you. Code requirements can change quickly, and every jurisdiction can have its own set of rules.

KEEP YOUR CONTRACTORS HONEST

Keep your contractors honest. If a permit and inspection are required for the work being done, insist that the contractors comply with the regulations. While it may not be your responsibility to see that contractors obtain the proper permits, you may be the one to suffer in the end. Make it your responsibility to supervise the acquisition of permits and the completion of required inspections.

How could a contractor's violation of code requirements hurt you? Serious violations could jeopardize your health and safety. For example, there are many plumbing-code requirements that exist for your protection. If an air gap is not installed with the drainage system of a dishwasher, it is possible that contaminated water in the drainage pipes will flow back into the dishwasher, creating a health hazard. If a plumber omits a required vent pipe, you may suffer from the effects of sewer gas. Water heaters that are installed without relief valves can blow up, causing death and destruction.

An electrician's failure to install a ground-fault interceptor could result in a fatal shock. Aside from the more deadly infractions, minor code violations can make your job less functional. If electrical outlets are not installed at the prescribed minimum distances, you may have an appliance with an electrical cord that is too short to reach an outlet.

If a heating technician installs heating units in the wrong locations, the performance of the heating system may not be satisfactory. All these possible code violations can present you with unwanted circumstances.

In addition to the health risks and inconveniences you might encounter with work done in a non-conforming manner, you might lose a substantial amount of money because of the violations. Suppose an electrician gave you a great price for rewiring your bathroom and did the job without a permit. What might happen? Here are some possibilities:

- The house could burn down.

- You could be electrocuted.

- An electrical inspector from the code enforcement-office might require all the bathroom walls to be torn open so the wiring could be inspected when the illegal job was discovered. That's right, the inspector could insist that you rip out your newly remodeled walls to expose all wiring that should have been inspected prior to concealment.

If you are forced to destroy your new job for the sake of an inspection that was never done, who is going to pay for the damage? It seems that the offending contractor should, but suppose the contractor has gone out of business or left town. You may be left holding the responsibility. Don't set yourself up for this type of disaster. Make sure all your work is done in compliance with code requirements.

10

Financing Your Project

What is involved in financing your remodeling effort? Can you just walk into your bank and ask for a loan and get it? Should you take out a personal loan to pay for the job, or should you refinance your house to cover the costs? These are only some of the questions that may come up before you begin work on your remodeling project.

Major remodeling jobs often cost more than $10,000, and this forces a lot of people to seek financing. Since the average person is not accustomed to working with financing, the task of finding and selecting the best loan can be an arduous one. Just as there are many types of products and designs to choose among for your job, there are also many forms of financing to consider.

DOES THE TYPE OF FINANCING YOU CHOOSE REALLY MAKE A DIFFERENCE?

Does the type of financing you choose really make a difference? It can make a big difference in the long-term cost of your job. The interest you pay on the loan will increase the overall cost of your remodeling work. If you have to pay a lender financing points to originate the loan, you will need more money to start the job, and the amortized interest will increase the long-term cost of the job.

> When you are counting on your remodeling loan to be tax-deductible, consult a tax professional before you accept the loan. Don't be fooled by a fast-talking lender. Confirm that the loan is a valid tax deduction for you before you accept the financing.

Your choice in loans can also affect how much you pay in income taxes. If you arrange the loan as a part of your home mortgage, the interest on the loan may be deductible. When the interest payments are deductible and result in less taxes to be paid, the job does not cost as much as it would with a loan where the interest was not tax-deductible.

If you are thinking of selling your home, the financing you choose could hamper the sale. Assume that your house has an assumable mortgage with an attractive interest rate. Selling the house with the assumable loan may give you a great advantage in a real-estate market with high interest rates. However, if you refinance the house at current interest rates to pay for a new kitchen, you have lost your edge in the sales market.

The selection of financing is a responsibility that should not be taken lightly. Not only is selecting the right financing important, but knowing how to apply for it can make a big difference in whether your loan is approved. Let's take a close look at how financing will play a role in your remodeling plans.

FIXED-RATE LOANS

Fixed-rate loans are the loans most people are best informed about. These loans operate on a simple principle: you borrow money at an agreed-upon interest rate and the rate never changes. The biggest advantage to this type of loan is that you know what your payments will be for every installment over the life of the loan. The biggest disadvantage to a fixed-rate loan is that because the interest rate is fixed, it is usually higher when the loan is obtained than rates for other types of loans are.

If you plan to keep your house for at least the next ten years and you are not much of a gambler, fixed-rate loans will probably suit you best. People who want to save money on the first few years of their loan and who are not afraid of a little risk will normally do better with an adjustable-rate loan.

Beware of balloon loans. These are loans where you may pay interest only for a period of years and then have the entire principal balance come due in one lump-sum payment. If your financial future doesn't work out the way you expect it to, a loan with a balloon payment can destroy you.

ADJUSTABLE-RATE LOANS

Adjustable-rate loans are common, and they offer many advantages. As the name implies, the interest rates for adjustable-rate loans can and usually do fluctuate. The rates typically go up rather than down, but they can go down. There are many variations of adjustable-rate loans; some are better than others.

Most adjustable-rate loans are adjusted annually, but others may adjust semi-annually or at other intervals. These loans are tied to specific indexes that affect how they are adjusted; treasury bills are one of the common indexes.

Since lenders are not making long-term commitments for interest rates on adjustable loans, the starting interest rates are often very attractive. It would not be uncommon to find adjustable-rate loans with starting interest rates several points lower than fixed-rate loans. This can save you money in the early years of the loan.

Stay away from loans that have negative amortization. You could wind up owning more money after a year's worth of payments than you owed when you started making payments.

Adjustable-rate loans that have annual and lifetime caps are safer than loans without caps. What are caps? The caps limit the amount of increase allowed in the interest rate of the loan. For example, if a loan has an annual cap of 2 percent, the interest rate cannot go up or down more than 2 percent in any given year. When the loan has a lifetime cap of 6 percent, the loan can never go up or down by more than 6 percent. For instance, if the loan had an interest rate of 6 percent when it was originated and the lifetime cap of the loan was 6 percent, the interest rate could never go higher than 12 percent.

> Loans without caps are dangerous and should be avoided; there are no limits on how high the rates for these loans can go.

There are so many types of adjustable-rate loans available that each loan must be studied carefully. Avoid loans that don't have annual and lifetime caps, and avoid loans that allow negative amortization. With negative amortization your payments can be very low, but after paying on the loan for five years you may owe more than you borrowed. Negative amortization allows interest that is not being paid (to keep the payments low) to accrue, resulting in a higher loan balance than you started with.

> If you plan to sell your house within five years of completing your remodeling project, an adjustable-rate loan could be a real money-saver.

IN-HOME FINANCING

Many contractors offer in-home financing plans. These plans are easy to apply for and are almost always approved if the applicant has equity in a home. There are many types of in-home financing plans. These plans typically charge a much higher interest rate than what a credit union or commercial bank would. However, points and closing costs are rarely an expensive factor with in-home financing.

An in-home financing contract is similar in many ways to the financing offered by car dealers. There are times when this type of financing is worth considering, but normally you can do better by dealing with your bank or credit union.

FINANCE COMPANIES

Finance companies love to make second mortgages to homeowners with plenty of equity in their homes. The rates and terms offered by financing companies fluctuate, but they are rarely as good as what is available from credit unions and banks.

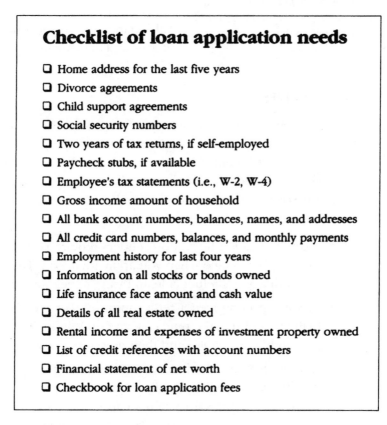

Checklist of loan application needs

❑ Home address for the last five years
❑ Divorce agreements
❑ Child support agreements
❑ Social security numbers
❑ Two years of tax returns, if self-employed
❑ Paycheck stubs, if available
❑ Employee's tax statements (i.e., W-2, W-4)
❑ Gross income amount of household
❑ All bank account numbers, balances, names, and addresses
❑ All credit card numbers, balances, and monthly payments
❑ Employment history for last four years
❑ Information on all stocks or bonds owned
❑ Life insurance face amount and cash value
❑ Details of all real estate owned
❑ Rental income and expenses of investment property owned
❑ List of credit references with account numbers
❑ Financial statement of net worth
❑ Checkbook for loan application fees

Figure 10.1 Loan application needs.

CONSULT YOUR ATTORNEY

Before you sign any financing agreement, it would be wise to consult your attorney. When you give a lender a mortgage against your home, the wrong words in the financing agreement could cause you to lose your house. While most banks use standard financing agreements, not all notes and mortgages are the same. When you are dealing with private financing from finance companies and in-home financing plans, you may run into some strange and undesirable language in the financing terms and agreement.

You will have to decide if you are willing to pay expensive legal rates to have potential financial agreements reviewed. Some homeowners choose to act on their own and save the money. What will you do? I would invest in a good attorney before I put my house on the line.

FINANCING FEES

Financing fees can add to the cost of your job. You know that the interest you pay on a loan over the coming years will increase the investment made in remodeling, but you may not be aware of some of the more immediate costs you may incur. Here are some examples of potential costs when obtaining financing:

- Loan application fees: Most lenders require you to pay a loan application fee when you apply for a loan. These fees are normally nonrefundable and vary in their amounts.

- Credit reporting fees: Credit reporting fees are also normally required at the time of loan application. These fees are rarely refundable, and their amounts also vary.

- Appraisal: When the loan you are applying for is a large one, the lender will normally require an appraisal of your property. The rules for when an appraisal is required vary from lender to lender, but if you are planning a complete bath-remodeling job, plan on the cost of an appraisal.

Appraisal fees vary from location to location and lender to lender, but budget a few hundred dollars for the fee. A phone call to your intended lender will give you solid numbers to put into your budget.

- Points: When talking about financing, points are fees equal to 1 percent of the loan amount. In other words, if the loan amount is $20,000 and there is a fee of two points, the fee will be $400. These points are sometimes called origination fees, discount points, and prepaid interest.

- Title searches: Title searches are normally required when a house is being used as security for a loan. These searches reveal any outstanding liens or encumbrances that affect a homeowner's equity position. Fees for title searches also vary, but expect them to be a few hundred dollars.

- Surveys: If your bathroom remodeling calls for adding space onto your home, expect to pay for a survey. A lender will want to be sure your addition is on your property and in compliance with zoning regulations. Surveys can cost only a few hundred dollars or they can cost much more, depending upon the size of your lot and the type of survey required by the lender.

- Other closing costs: There can be other closing costs that will have to be paid when you borrow for remodeling. These costs might include legal fees, filing fees, and other types of fees. Any reputable lender will provide you with an estimate of what your closing costs will be before you commit to a loan.

HOW MUCH WILL BORROWING MONEY ADD TO THE COST OF YOUR JOB?

How much will borrowing money add to the cost of your job? It will depend on where you borrow the money and what the individual lender charges, as well as what the interest rates are.

You have just seen what some of the hidden costs might amount to, so now let's see how much the interest on a loan could cost you.

Assume that you are doing a major remodeling job and that the amount of money borrowed from your bank will be $20,000. The interest rate is 10 percent, and the loan is a fixed-rate loan with a term of fifteen years. This means that you will make monthly payments of about $215 for 180 months. The total amount of these payments will be $38,700, nearly twice what you originally paid for the job. These kinds of numbers can be real eye-openers. Granted, a 10-percent rate is high today, but rates are likely to go up and this is merely an example.

Now let's look at the same job but with the money being borrowed from an in-home financing plan. The interest rate on this financing is 11.5 percent, and the rest of the terms are the same. Your payments for this loan will be about $234 a month. This doesn't sound like a big difference, but let's see what the total cost of the job is with this financing. The total amount of these payments will be $42,120, $3,420 more than with the bank financing.

> Before you borrow money, make sure you understand what you are signing and the monthly and total cost of your financing. This is one area of remodeling work where you don't want any surprises.

MAKING LOAN APPLICATIONS

Once you have decided to finance your job, you have to prepare for making a loan application. This is not a complicated process, but it helps to have all your documents in order. There will be many items a lender wants before approving your loan.

Getting a personal loan is a little different than getting a loan based on the value of the home improvements. You will need less documentation for a personal loan. As we look at your needs for a successful loan application, we will examine them on the assumption that the home improvements will be a factor in the loan approval. If you have plenty of equity in your home or a strong line of credit, you will not need all the items about to be discussed. Here are some of the requirements that you might expect when applying for financing:

- Plans and specifications: If a loan is being based on the value of an intended home improvement, the lender will want plans and specifications. He or she will review your plans and specs and will probably have a before-and-after appraisal done on your home, based on the plans and specs you provide.

- Permits: Some lenders require all necessary permits to be purchased prior to closing a home-improvement loan. This assures the lender that the work may be done with the necessary code approvals.

- Tax returns: Some lenders want to see tax returns for the last two years. This normally is not required unless you are self-employed.

- Bank accounts: You will be asked to list all your bank accounts. Account numbers will be needed, and you must identify the type of accounts you have, such as checking and savings accounts.

- Financial statements: While you will not need a formal financial statement unless you are self-employed, you will have to list all your assets and liabilities. The liabilities will include credit-card debt, car loans, school loans, and any other financial obligations you have. Account numbers will be needed for each loan you have outstanding.

- Social Security numbers: Social Security numbers are required on loan applications, so if you don't know yours, find out what it is and bring it to loan application with you.

- Employment history: The loan application will request information about your employment history. If you have had different jobs in the last five years, be prepared to list the dates of your employment, the locations, your earnings, your supervisor, and the employers' addresses.

- Previous addresses: If you have lived at other addresses in the past few years, be prepared to provide detailed information.

- Credit references: There will be space on the loan application for listing credit references. When you list these references, you should be prepared to provide account numbers and addresses.

Figure 10.2 Financing can help achieve designs that you never thought were possible. *Courtesy of Wellborn Cabinet, Inc.*

There may be a need for other information. For example, if you have been divorced, you may be required to provide a copy of the divorce decree. If you have had credit problems in the past, be prepared to explain in writing why you had trouble maintaining good credit.

LOAN APPROVAL

Once formal loan application is made, about all you can do is wait. The process can take as little as a week or as long as two months. You will normally be notified by mail when your loan is approved or denied.

Once the loan is approved, a closing date will be arranged. This will be the day you sign all the papers and receive the money to begin your project. Loan closings are usually simple procedures that take less than an hour to complete. However, you may wish to take your attorney to the closing with you. The papers you sign may affect your position of ownership in your home.

Now that all the preliminary work is out of the way, let's move on to the next chapter and get involved with the remodeling work.

11

Ripping Out Baths

The demolition process involved with remodeling a bathroom can be a major undertaking. Depending upon existing conditions and the degree of demolition needed, the job can consume days of your time. Not only can the demolition process be time-consuming, it can be dangerous for both you and your house. If you don't follow proper procedures, you may electrocute yourself or flood your home with water.

Should you do your own demolition work? If you are in good health and are handy, you shouldn't have any trouble completing your own demolition work. However, you should have an understanding for what is involved in the rip-out process before you begin gutting your bathroom. Here are some considerations to be aware of:

- There are safety hazards involved with demolition work.
- Eye protection should be worn at all times during the rip-out phase.
- Ear protection may be needed for some parts of a job.

- The proper clothing and footwear can reduce cuts, scratches, and punctures.

- There is a good chance some of your demolition work will be done while standing on a ladder, so caution must be exercised to avoid falling.

This book is not intended to teach you on-the-job safety procedures. There simply isn't room to discuss proper lifting procedures and other safety-related topics. But it is important that you know that there are dangers involved with doing remodeling work. These dangers extend throughout a job. If you are not aware of how to work safely with tools, ladders, and general remodeling work, consult books on the subject of safety.

Don't be too quick to start tearing a room apart. There is planning and preparation work needed before the actual work of ripping out a room should commence.

While I am not going to attempt to teach you safety procedures, I will show you some tricks of the trade. My many years of experience have allowed me to develop some techniques to make remodeling work easier, and I'm going to share these secrets with you. If you are ready to get down to some serious work, roll up your sleeves, and we will get started.

PREPARING FOR DEMOLITION

By preparing for demolition work in advance you can avoid a number of problems. One of the first problems inexperienced people run into with demolition work is the mess it makes in the rest of the home. There is a lot of dust and debris involved with demolition work, and keeping the mess contained in the room being ripped out is the first order of business.

Disposing of Debris

Before you can begin the containment process, you must have a plan for disposing of the debris you will create. If the room

Figure 11.1   Even though the demolition will be messy, the end result is worth the trouble. *Courtesy of Moen, Inc.*

being remodeled has a window or door that opens to the outside, you may be able to place a trash container near the opening and toss the debris out as you go along. This not only controls clutter in the workplace; it also makes the job go faster. If you have to pile the rubbish in the room and them haul it out to a trash container, you are handling the materials twice. Try to position a trash container where it is easily accessible and dispose of your refuse as you create it.

If you are working in a room that is on a second or third floor of your home, you may want to build a chute for your trash removal. Another option is to buy a tube that is used for pier foundations. This tube can be used as an enclosed chute and is relatively inexpensive. Set a trash container below the upstairs window where you will be discarding debris. Use framing lumber and plywood to build a trash chute if your job is not compatible with the cylindrical tube. The chute should have side rails that prevent debris from falling over the sides as it slides down to the trash container. The chute will resemble a sliding board. Have the chute extend from the trash container to the window and secure it firmly. With the chute in place, you can dump debris out the upstairs window and have it land safely in the container.

> Using a debris chute or tube will reduce the amount of time and effort spent in getting debris out of the project room and into the trash pile or container.

Dust Containment

Dust containment will be your next step in preparing for demolition. If you don't seal off the room you are remodeling from the rest of your home, dust will find its way into your carpeting and all over your home. The work involved in setting up dust containment is much easier than trying to deal with dust all over the home.

All you need for dust containment is some plastic and some duct tape. Seal all doors and other openings between the workspace and the remainder of the home with sheets of plastic. Cut the sheets larger than the openings you are covering and allow the plastic to extend several inches past the frame of the opening. Duct tape can be used to affix the plastic to the walls of the room you are working in. Keep the plastic

Figure 11.2   An example of a finished bathroom. *Courtesy of Armstrong*

on the side of the opening where you will be working, not on the side where the rest of your home is. Tape the plastic to the walls and floor using long strips of tape. Don't leave any portion of the seams untaped.

If you must use a door that opens into other living space for access to the room being remodeled, you may want to use an alternative method for sealing the opening. Since pulling tape loose from the walls and floor every time you want to enter or exit a room can be annoying, you might want to get a bit more creative.

To avoid some hassles with coming and going, cut the sheet of plastic to be used for covering an entrance extra large. Tape the top and one side of the plastic to the walls in the way described above. Attach a 2 x 4 to the bottom of the plastic and roll the plastic around the piece of wood until the vertical fit is tight. The weight of the wood will hold the plastic down and eliminate the need for taping it to the floor.

On the remaining side of the opening, tape the plastic to the wall at the top, middle, and bottom, but don't use long strips of tape; small pieces will do fine. There will be gaps along this edge of the plastic. Left alone, these gaps will allow dust to escape the room. To remedy the dust situation without sealing yourself in tightly, hang a second piece of plastic to overlap the lightly taped edge.

The second sheet of plastic should be taped to the wall at the top and along one edge with long strips of tape. The section of plastic that overlaps the other plastic should be taped at the top, middle, and bottom with the use of minimal tape. When this is done, the room is sealed, but you can come and go easily. All you have to do to open the exit is pull the tape from the bottom and middle section of the overlap and inner plastic. The bottom of the main covering will move easily when you push the wood to one side. This type of arrangement is effective in controlling dust while allowing reasonable ingress and egress to the space.

> If you have sensitive flooring, such as finished hardwood floors, cover them with construction paper to protect the finish. Failing to do this could result in scratches and scars if dust settles on the surface of the flooring.

RIPPING OUT A BATHROOM

What is involved in ripping out a bathroom? If you are taking the room down to bare studs and subfloor, there is a lot of work involved. You will have to work with plumbing and electrical devices, and you may have to work with part of your heating system. These mechanical systems must be treated with respect. The job will also involve removing wall coverings, ceilings, and floor coverings. While most homeowners have little problem with the demolition of a bathroom, there are some things to be careful of. Let's look at how the demolition process might go in an average bathroom.

Assume that you have your bathroom prepared for demolition and you are equipped with the proper safety precautions:

eye protection and so on. In this job none of the existing fixtures will be salvaged. Where will you start the rip-out? The logical place to begin is with the plumbing fixtures.

Plumbing

Before you begin tearing out your old plumbing, make sure the water supply to the fixtures is cut off. Don't assume that a closed valve means that the water is off. Some valves fail with age. After you have closed the appropriate valves, test each fixture to see that the water is in fact shut off.

Begin by removing the toilet. A screwdriver and an adjustable wrench are the only tools you should need for this part of the job. Flush the toilet to evacuate most of the water. Remove the nuts on the bolts extending through the base of the toilet. If the nuts are seized and will not turn, they can be cut off with a hacksaw blade. Loosen the nut that secures the water supply to the toilet tank. The toilet can now be lifted off the floor and removed.

A complete toilet can be awkward and heavy for inexperienced people to handle. In most cases the tank of the toilet can be separated from the bowl, making removal easier on your back. Toilet tanks that are removable are attached with brass bolts and nuts. By putting a screwdriver in the head of the bolt and turning the nuts, the tank should be easy to remove. Sometimes the bolts are stubborn, and a hacksaw blade is needed to cut them. If you get the bright idea to break the toilet into pieces with a hammer, be advised that the broken china can inflict nasty wounds when shattered.

> Taking a hammer to a toilet or lavatory can be dangerous. The china will break and fly about the room. Without protective clothing and equipment, you can be cut by the flying china. While it is common to use a sledgehammer to break up cast-iron tubs, it is not advisable to break toilets and lavatories. In any type of demo work, you must be dressed properly and have adequate protection for your eyes, ears, hands, and face.

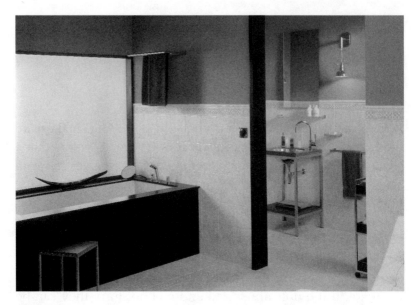

Figure 11.3   An example of a very unusual bathroom design. *Courtesy of Armstrong*

With the toilet out of the way, you are ready to move on to the lavatory. Disconnect the trap from the fixture. This can usually be done with a set of wide-jaw pliers. The water supplies must also be disconnected. You should be able to loosen the small compression nuts that hold the supply tubes into the cut-off valves without any trouble. Once the waste and water lines are loose, you can remove the basin.

If you have a wall-hung lavatory, it should lift straight up off its wall bracket. Some wall-hung lavatories are secured to the wall with lag bolts. If you cannot lift the bowl off the bracket, look for bolts securing it to the wall. If none are found, exert some extra pressure to remove the lavatory. You may have to twist it left and right to get the bowl off the bracket.

When your bathroom is equipped with a vanity and top, the removal process is different. The vanity top is probably attached to the wall with caulking. Run a knife along the joints where the top meets the wall. Look to see if the top is attached to the vanity or simply sitting on it. Remove any screws holding the top to the cabinet; it should lift off. Before trying to remove the base cabinet, check to see if it is screwed to the wall.

With the lavatory and toilet out of the way, you have more space to work with in removing the bathtub. Before you can remove the bathtub, you must remove the walls that overlap its edges. Strip the walls surrounding the bathtub to reveal the tub edges.

> Don't use a saw to cut into walls. There may be live electrical wires in the wall. Beat a hole in the wall with a hammer to inspect for wiring before using a saw on the wall.

A hammer works well in removing the walls. Don't cut into the walls blindly with a saw. You might hit live electrical wires. Beat holes in the walls with a hammer and pull the wall covering off. If you have your heart set on using a saw, at least open the walls with a hammer and check for wiring and plumbing before using it.

When you have the walls around the tub stripped to the bare studs, you can start the removal process for the tub. Remove the tub faucet first, but be sure the water to the pipes is turned off. Unlike the other fixtures, the faucet for the bathtub will not have small supply tubes; it will be connected directly to half-inch tubing or pipe. There may be unions in the pipes to make removal easy, but you will probably have to cut the pipes. This can be done with a hacksaw blade or pipe cutters.

The next step is the removal of the tub waste and overflow. From inside the bathtub, remove the screws holding the trim on the overflow. If the drain has a strainer on it, remove the screw securing the strainer and expose the crossbars in the drain. Insert two thick-shaft screwdrivers into the drain and cross the shafts. By creating an "X" with the screwdrivers you will be able to loosen and remove the drain. Turn the drain counterclockwise to unscrew it.

When all the plumbing connections are loose, you can remove the tub. If the bathtub is a one-piece tub-shower combination, it should be secured to the stud walls with nails or screws. Once the tub is free of the walls, you will probably have to cut it into sections. One-piece tub-shower units will not normally fit through interior doors or travel up or down finished stairs. You can cut the unit into pieces with a hacksaw blade, but the job goes much more quickly with a reciprocating saw.

Standard bathtubs (without integral shower walls) are not normally attached to the stud walls; they sit on supports. All that is required to remove a standard tub is to lift and slide it out of the opening. This sounds easier than it sometimes is.

Plastic, fiberglass, and steel bathtubs can be removed without too much strain, but cast-iron tubs are another matter entirely. Cast-iron tubs can weigh over 400 pounds, and wrestling one out of its resting place can be very difficult, even

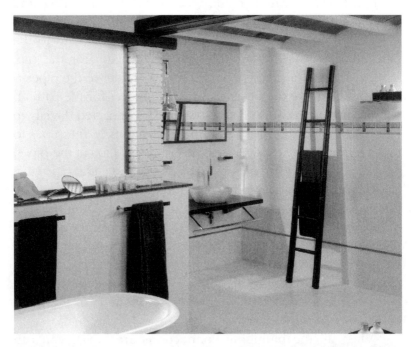

Figure 11.4   Another exotic bathroom design. *Courtesy of Armstrong*

for seasoned professionals. Many professionals use sledgehammers to break cast-iron tubs into manageable pieces. If you do this, wear eye and ear protection, along with clothing that will protect you from sharp pieces of the tub that may fly into your body.

Heating

You may not have to do much with the heating system in your bathroom. If your heat comes in through ducts in the floor, all you have to do is remove the register from the duct and protect the open duct from falling debris. You can stuff a towel in the duct or cut a piece of plywood to cover the opening.

If you have hot-water baseboard heat, you will want to remove the baseboard-heating element. This will require shutting down the boiler and may require draining the heating system. If you are working on the top floor of your home, you shouldn't need to drain much water from the heating system before cutting the supply and return pipes at the baseboard unit. However, if there is heat installed in rooms above your bathroom, drain the heating system to a point below the bathroom.

There will be removable end-caps on the baseboard-heating unit. Remove these caps by pulling them off to reveal the supply and return pipes. The pipes should be copper. If they are, they can be cut with a hacksaw or a pipe cutter. Once the pipes are cut, remove the screws that hold the baseboard unit to the wall and remove the heating unit.

> If you have an old house that is equipped with radiators, try to avoid removing them. Old radiators can become damaged when moved, and they are hard and expensive to replace.

Electrical

Now that the plumbing and heating are out of the way, you are ready to work with electrical devices and fixtures. Turn off the power to the bathroom. If you are going to remove your own electrical fixtures, be certain the electricity to the wires is off. Use an electrical meter to test each wire before working with it. If you don't know how to use an electrical meter, you have no business working with electricity; call in a professional.

With the power turned off, remove all cover plates from switches and outlets. Remove the globes or shades on your lights and the light bulbs. Most electrical fixtures are attached to their electrical boxes with threaded rod and nuts. Remove these nuts and the fixture should come loose. Remove the wire nuts (plastic covers protecting the wires) and test for electricity.

> Don't attempt to work with electricity if you are not absolutely certain that you have the ability to do so. One mistake with electrical work can be the last one you will ever make in life.

When you are sure the power is off, separate the fixture wires from the house wiring. Install wire nuts on the house wiring and tuck it back into the electrical box.

If you have electric baseboard heat, it should be attached to the wall with screws. Before handling the wiring to the heat, make sure the electricity is off. You cannot assume that all your bathroom wiring is on the same circuit. Just because the light doesn't have power coming to it doesn't mean the heat is safe to work with.

Walls and Ceilings

Removing finished walls and ceilings is not difficult if you are working with drywall. Use a hammer to open the walls and ceiling and to expose all wiring, plumbing, and heating. A dust mask will help protect you from the massive amounts of dust this process will create. You can then either continue to demo the walls and ceiling with a hammer or you can cut the drywall out with a saw. Window and door trim will also have to be removed during this stage.

If your walls are made of plaster, you have a lot more work in front of you. A reciprocating saw is the fastest way to cut through plaster and the lathe behind it. However, use a hammer to open sections of the wall before running the saw through the plaster. It is easy to cut wires and plumbing by accident.

Flooring

Removing vinyl flooring is not difficult. Start by removing all baseboard trim and shoe molding. When the molding is removed, the edges of the flooring will be exposed. You may be able to grasp the ends and pull the flooring up. If the floor is difficult to remove, you can use a floor scraper to remove it.

If you are removing a ceramic tile floor, you can chisel the tiles up or break them out with a hammer. Remember to protect your eyes and body from the sharp slivers created by breaking the tile.

Odds and Ends

There will be some odds and ends to tend to. Go around the bathroom and remove all existing nails that protrude from the walls and ceiling. Sweep the floor and scrape it until it is clean. Cap all pipes to keep debris from entering them. Make sure all electrical wires are protected with wire nuts. Look around and tidy up any loose ends.

When you are involved with remodeling you are sure to run into some unexpected conditions. You might find that your subfloor and even your floor joists have been damaged by water leaks. When you open a wall, you might come face to face with a nest of angry bees. There are all sorts of things that can disrupt your remodeling plans. You have to prepare for any number of unexpected events. This is what keeps the work exciting.

12

Unexpected Conditions

It is very likely that you will run into some unexpected existing conditions when you begin remodeling your bathroom. You can plan and plan for a perfect job, but you cannot always plan for the unexpected. However, if you know what types of unexpected conditions may exist, you can hedge your odds for success.

Professionals are sometimes plagued with problems with existing conditions. It might be that the water pipes to a lavatory froze at some time and swelled to a point where fittings will not slide over them. The problem could be a wall that is not plumb or a floor that is not level. When you are installing vanity cabinets, it is very important to have level floors and plumb walls to work with.

While most of the unexpected problems that arise have to do with building materials and conditions, it is not unusual for other types of problems to pop up from time to time. When you open an existing wall, you never know what you might find. There could be a nest of bees hiding behind the drywall or an angry rodent. If you remove the ceiling in a bathroom that has its own attic, you may be surprised to see a swarm of bats flying about, startled by your actions. Crawling under the house to work on the plumbing might bring you face to face with a snake or feral cat. Without doubt there are plenty of opportunities for surprises when remodeling.

Some of the wildlife stories you have just heard about may be difficult to believe, but I have had all these things happen to me. I have run into skunks, rattlesnakes, rats, squirrels, feral cats, bats, bees, porcupines, and other types of wildlife during my remodeling career. Most of the encounters ended amicably, but some of them were exciting for a while. I tell you this not to scare you but to prepare you. If you are aware of what you might find at different stages of your job, you will be better prepared to deal with the problems.

To elaborate on what you might expect during your remodeling experience, let's look at the various phases of your job and see what might get in your way.

FLOORS

When you are working with your floors, you may find that the subfloor or floor joists must be replaced. The damaged floor structure may not be evident until you have removed the finish floor covering. Water damage is a frequent cause of this type of problem, but it could result from termites or other wood-infesting insects. The floor joists may simply have rotted to a point where they are no longer structurally sound.

WALLS AND CEILINGS

Walls and ceilings can conceal all types of potential problems. When the area cavities behind finished walls and ceilings are exposed, you may find that the insulation has become dilapidated or was never installed. This can add to your work with unexpected labor and material costs.

Bees and snakes often take up residency in hollow walls. It can be very unnerving to cut out a wall section and be swarmed by angry bees, and seeing a big snake slither out of sight in the wall can ruin your whole day. You never know when the snake will turn up again.

The walls studs could be covered with the telltale sign of termites. Not only will you have damaged studs to replace, you will be faced with the expense of professional termite treatments.

> When removing a ceiling, protect yourself from falling debris. It is common for loose-fill insulation to come pouring out of ceilings that have attic space above them. Eye protection and a dust mask will help to prevent complications from this type of work.

The bottom plate of your wall may have rotted, causing much more extensive work to the wall than you expected.

Removing a ceiling can reveal water damage to the floor above you. There are many times when leaking plumbing fixtures and water damage go undetected until a ceiling is removed.

Removing a ceiling that has an attic above it can cover you in insulation. Most attics have insula-

> Attics can be home to a number of wild inhabitants. Bats, birds, snakes, rats, mice, squirrels, and even raccoons can be found in attics. Imagine your surprise if you are dropping a ceiling and confront a curious raccoon. If the ceiling you are working with has an attic above it, attempt to inspect the attic prior to opening the ceiling.

tion lying between the joists on the back side of the ceiling you are working with. Cutting out a section of this type of ceiling and getting a face full of loose insulation is no fun.

MECHANICAL SYSTEMS

Mechanical systems can cause much frustration. You may think you are working with copper plumbing pipes only to discover that they are brass. Old brass pipe can be cut with standard copper cutters, but copper fittings will not normally slide over the brass pipe. You might find that the electrical wiring in your bathroom was run in a haphazard way and that some wires are dead while others are hot. This can be a shocking experience.

Most mechanical problems can be overcome with minimal trouble, but some of them are serious. For example, you may discover that your toilet was not mounted to a flange and that the pipe connecting to the toilet was an old-fashioned lead bend. If this is the case, your main drainage system is old enough to be made from cast-iron pipe, and converting it to plastic pipe will require a special soil pipe cutter.

If you relocate a radiator that provides heat to your room, you may find that it leaks when you reinstall it. This can be expensive to correct, not to mention frustrating.

CABINETS

Installing cabinets is very difficult if you do not have level floors and plumb walls to work with. You can spend many hours installing shims to get the cabinets to fit properly. If you don't take the time to level the cabinets, the doors and drawers are not likely to work smoothly.

Check the walls early on in your job to see that they are plumb. Walls that are not plumb will result in gaps where counters or cabinets sit against the walls. It is far better to deal with walls that are out of plumb before you attempt to make new installations.

COUNTERTOPS

When you install your new countertop, you may discover that your walls are very much out of plumb. The top will be tight against the wall at one point but have a gap at another. If you don't know that the wall is out of plumb until you install the countertop, the cost of your job will go up considerably. If the gap is significant, you will have to remove the finished wall covering and use furring strips to build the wall out or order a new countertop. Either way you are going to lose time and money.

UNDER THE HOUSE

If you have to crawl under your house during your project, you may find any number of unexpected problems. Your floor joists could be riddled with holes from powder post beetles. Water could be standing under the home, creating mold and rotting your wood. While you are under your home, look for any defects that might exist. You may not want to fix them, but it is better to fix them before the problem escalates.

There is always the chance that animals will be seeking shelter in your crawl space, so don't go under the house without proper lighting.

After you have successfully completed the demolition phase of your remodeling, it is time to start putting it all back together again. Unlike trying to put Humpty Dumpty back together again, you won't need all the King's men and all the King's horses to put your bathroom back together. You may need some help from friends or contractors, but the job is manageable.

Anticipating what you are doing and thinking are the best ways to avoid serious problems. If you consider what could be inside a wall or under a floor, you are less likely to experience trouble with your remodeling. For example, failure to think about wires that are installed in a wall could cause you to cut them with a saw. This could result in an electrical shock, and it will certainly result in additional work. Think before you act, and you will have a good chance of getting through your job with minimal interference.

The task of restoring your bath will go much more smoothly if you do it in an organized manner. There will be a lot of work to do, and some of the work will be much more difficult if it is not done in the right sequence. For example, should you install your countertop before or after you paint your walls? Most contractors paint first and then install countertops.

Why do they do it in this order? If the counter is installed first, there will be no paint on the wall behind the counter. This means the painter will have to paint right up to the edge of the counter. Painting the room will be much more difficult with the cabinets and countertops installed. Access to the walls will not be as good, and there is always a risk of spilling paint on expensive cabinets and counters.

Some professionals and many homeowners believe countertops should be installed before a room is painted. Why? Because they think that freshly painted walls are likely to be dinged when the counters are installed. It is probable that some scuffing will occur on the walls, but it is much easier to touch up a few dings with the counter in place than it is to paint all the walls with the counter in the way. This is only one example of how the proper planning and scheduling can make your job run more smoothly. Let's look ahead now to some of the other ways you can improve your efficiency in remodeling.

CREATE A PRODUCTION SCHEDULE

Before you begin your work, create a production schedule. Organize all the work that will need to be done into a logical progression. Break the work down into phases, such as rough-in plumbing and final plumbing. It is unlikely that your schedule will work out exactly the way you design it, but by having it in writing you will be less likely to overlook work that needs to be done, and you have a better chance of keeping the job on track.

When you have tentative dates by all your work phases, note the work that will be done for you by contractors.

Once you have a list of all the types of work required in your job, assign dates to each task. For example, you might schedule rough plumbing for June 10 and rough electrical work for June 11. Remember to allow time for code inspections. Some phases of your job will not be able to progress until other phases are inspected and approved. For example, you could not hang drywall and conceal rough plumbing and electrical work until inspection approval has been given.

This will enable you to give them plenty of advance notice of when their services will be required. If part of the job doesn't go according to schedule, remember to change the dates for all aspects of the work that will be affected.

It may take several attempts before you have revised your production schedule to a point where it is reasonably accurate. If you will be using contractors, ask them to provide estimates for the amount of time they will need to complete their work.

It is unlikely that you will know how much time to allow for the various types of work. If you are doing all the work yourself, the exact timing will not be as critical as it might with contractors. However, even if you are doing the work yourself, you must develop a rough idea of when various phases will be complete. Otherwise you will not be able to project the dates for ordering cabinets, countertops, and other supplies.

LAY YOUR PRODUCTION SCHEDULE OUT IN LOGICAL ORDER

It is important that you lay your production schedule out in logical order. Since you are not an experienced general contractor or remodeler, you are not likely to know what a logical order is. Don't worry—you are about to find out. All time estimates are based on professional work. You should adjust the times to compensate for your skill levels.

Preparation

Preparation work is the first phase of any remodeling job. This is when you make arrangements for trash removal and dust control. Permits are often obtained at this

Check with your suppliers to determine how much advance notice is required to have various items delivered when you want them.

stage of the job. The preparation work can usually be completed in a day.

Demolition

Demolition work is the next step in remodeling a bathroom. However, before you begin tearing out what you have, order the materials to put the space back together again. Check with your suppliers to determine how much advance notice is required to have various items delivered when you want them. For example, you can probably get a load of lumber in a day or two, but it could take weeks to get new cabinets. Demolition work for a bathroom can often be done in one day.

Rough Framing

If you are relocating or adding walls in your remodeling job, now is the time to do it. Framing work often requires a building permit and an inspection. If you are planning to install underlayment on your subfloor, do it while you are in the rough-framing phase. Most framing for interior bathroom remodeling can be completed in a day or two.

Plumbing

Once the demo work is done, it is time to make any adjustments needed in the rough plumbing. If you are installing a new bathtub or shower, this is the time to do it. This work usually requires a permit and inspection. The rough plumbing for a bathroom rarely takes more than a day to complete.

Heating

Heating work is normally done after plumbing and prior to electrical work. You may need a permit and inspection for the heating work required in your job. The heating work required for a typical bathroom remodeling job will take no more than a day.

Electrical

Electrical work typically follows plumbing and heating work. Again, a permit and inspection may be required. This work should be completed in one day.

Insulation

If you have any insulation work to be done, it should be done after the plumbing, heating, and electrical work has been done and inspected. Many code jurisdictions require insulation to be inspected prior to concealment. The building permit obtained for the job covers insulation work. Installing insulation doesn't take long, but allow a day for it in your production schedule. Jobs don't run as smoothly when different trades are on the job on the same day.

Drywall

Once all your mechanical work and insulation are installed and inspected, you are ready for drywall. This work is covered under the building permit, but it often requires an inspection. Rarely will there be a specific drywall inspection, but the work will be examined when the final building inspection is done.

After the drywall is hung and taped, it will need to be finished. This is a lengthy process, since it requires days during which you will be unable to do much of anything else. Three coats of joint compound are typically installed on new drywall seams, corners, and dimples. Each coat of compound must dry and be sanded before the next layer is applied. The sanding is very dusty work.

> Before drywall seams are sanded, be sure that you have dust-containment procedures in place. The fine dust from the sanding will invade all the nooks and crannies of your home without proper containment.

Drywall can be hung on the walls and ceiling of a room in a day. The taping and first coat of compound can also be installed on the first day. By the second day the second coat of compound can usually be applied, and by the beginning of the fourth day the walls and ceiling should be ready for paint

Painting

The next phase of work to have done is the painting. Generally at least a primer coat and one coat of paint are used, and this work will take at least two days.

Finish Flooring

Finish flooring is normally done next. This part of the job doesn't require a special inspection, and it should be completed in a day.

Protect new flooring once it is installed. Some contractors apply a layer or two of construction paper for protection. When possible it can be desirable to cover the flooring with cardboard or some other protective material. A favorite of mine is a thin, inexpensive carpet pad. I place the pad over the new flooring and then seal the area with plastic or paper.

Cabinets, Counters, and Fixtures

The cabinets, counters, and fixtures can be installed after the finish flooring is in. Individual inspections are not required for this work. These items will be inspected when all final inspections are done.

Professional plumbers can set all the fixtures for a bathroom in less than a day. Electricians and heating mechanics can be in and out in a day when setting fixtures.

Trim Work

Once all the cabinets and fixtures are in, the trim work can be done. Some people install trim before fixtures and cabinets are installed, and others do it afterwards. Either way will work, but I prefer to have the trim installed afterwards; it always seems to fit better that way.

Most professionals paint or stain their trim before they install it. Then all that is required once the trim is in place is filling nail holes and touch-up work on the paint or stain. Installing trim can be time-consuming, but any good professional can trim a bath in a day.

Figure 12.1   An example of an expertly-finished bathroom. *Courtesy of Armstrong*

Final Touches

There be many small final touches to do before the job can be considered finished. These odds and ends can take a day or two to complete.

Cleaning Up

Cleaning up the mess and the new fixtures will be the last phase in the work. This should take less than a day.

Final Inspections

Once the job is completely finished, you are ready for your final inspections. This may involve a visit from a plumbing inspector, electrical inspector, HVAC inspector, and building inspector.

Some jobs will require work not mentioned in this outline, such as tile work or decorative stenciling. This list of work should give you a good idea of how to plan your production schedule. There is little doubt that you will have to make adjustments for the time allocations, but with some advance planning and effort you can make your job run smoother with a production schedule.

13

Flooring

T he floor of your bathroom is a focal point of the room. The right floor coverings can distinguish the room and make a statement about its owner. The wrong floor covering can darken the room or give an unbalanced appearance. Your choice in a finished floor covering will have a strong effect on the overall appearance of your remodeled room.

Even with the best-finished floor covering available, bad subflooring and floor joists can prohibit your room from being outstanding. If the floor squeaks every time you walk across it, you will notice the squeak more than you will the attractive floor covering. If the joists are weak and the floor is spongy, you may wonder when your toilet is going to fall through the floor.

Most of the work done in routine bathroom remodeling is not of a structural nature, but the flooring is. While you don't need superior framing skills to build a linen closet, you may need them to repair damaged floor joists. It is not common to discover rotted or damaged floor joists, but it does happen. What would you do if you removed water-stained

Figure 13.1   Flooring can make the difference in the beauty of your bathroom. *Courtesy of Armstrong*

subflooring and found the tops of three floor joists rotted to a point where the point of a pencil would penetrate them? A lot of people would panic and assume the joist would have to be removed and replaced. It is possible that the joists would need replacement, but it is more likely that a little repair job could solve the problem.

This chapter is going to show you many aspects of your flooring system. You will learn about floor joists, subflooring, underlayment, and finished floor coverings. There will be tips on how to handle rotten joists the easy way, and how to get the bubbles out of your new vinyl flooring. If you are ready, let's get on with an in-depth study of floors.

FLOOR JOISTS

Floor joists are the structural members that support the sub-floor. They are normally boards with dimensions ranging from 2 x 8 inches to 2 x 12 inches.

> Never skimp on the size of floor joists. If you are dealing with a span that is a borderline case between the minimum size required and the next larger, go with the larger joist.

Floor joists span the distances between outside walls and support girders. The length of the span and the use of the floor influence the size of the joist.

If you remove your subfloor and find a few of the floor joists to be rotted, you may not have to replace them. It is entirely possible that you can add new supports without removing the old ones. In many cases all you will have to do is slide new joists in place on each side of the damaged one and nail them to it. The new joists should be the same length and have the same dimensions as the old joist. This is normally not a very difficult procedure and it works fine in most circumstances.

If only a small section of a joist is damaged, you may be able to get by with scabbing new pieces of wood on the old joist. Assume that an existing joist was damaged but only for about 2 feet of its length. You may be able to attach new wood on each side of the damaged area to avoid installing complete joists. The scab wood should extend well past the damaged area: in this case about 4 feet in length should be sufficient. This practice may not be suitable in all situations. Check with your local building inspector before relying on this type of repair.

Another option available for joists with small areas of damage is a process in which the old joist is headed off. To head off a joist, you cut out the damaged section. Then joist-size material is used to span the distance between sound joists. The cut end of the damaged joist is attached to the new wood that is running perpendicular to the cut end. This procedure is also used when an opening is needed between floor joists for the passage of chimneys or stairs.

SUBFLOORING

Subflooring is the flooring that attaches to floor joists. It can be made of boards, but it is usually made with sheets of plywood or particleboard. Most jurisdictions allow two options when installing subflooring. Either one layer of tongue-and-groove material or two layers of standard plywood or particleboard may be used. It is not normally acceptable to install a single layer of material that is not fitted with a tongue-and-groove installation.

UNDERLAYMENT

Underlayment is usually a thin (about ¼ inch) sheet of plywood that is laid over a subfloor. The underlayment is normally sanded on one side and provides a smooth surface for the installation of finish flooring.

VINYL FLOORING

Vinyl flooring is the most common type of flooring used in kitchens and bathrooms. It is generally available in widths of 6 or 12 feet. Vinyl flooring can be tricky to install, but the job can be done by anyone with average skills and patience.

Before you start installing your new vinyl, make sure the area of the installation is clean and smooth. The surface should be flat and without cracks, depressions, or bulges. Cracks

> Don't attempt to install vinyl flooring when it is cold. The material will be difficult to work with, and it may be brittle.

Figure 13.2   Examples of vinyl flooring. *Courtesy of Armstrong*

in floors can be filled with special compounds. These filling compounds are available from the same stores that sell flooring.

Before installing your floor, roll the flooring up with the finish side facing outward. Leave it in this position for a full day. Maintain an even temperature of about 65 degrees F. in the room where the vinyl is being stored.

Vinyl flooring should be laid out in a room with enough excess vinyl that the flooring rolls up on the walls. A utility knife is one of the best tools to cut vinyl flooring with.

When you are securing your flooring to the subfloor, you may use adhesive, tape, staples, or a combination.

If your floor is going to require seams, make them before installing the flooring. Lay two pieces of flooring in place so that they overlap. Make sure the pattern meets and matches. Using a straightedge and a utility knife, cut through both pieces of flooring where the seam will be made. Remove the scrap flooring and attach the two pieces of finish flooring to the floor at the seam. Use a hand roller to press the flooring down. The back of the flooring should be in contact with the adhesive or tape you are using for the installation. Cover the seam with a sealing compound.

A floor roller should be used to roll wrinkles out of a new installation. Rollers can be rented at tool centers. When the vinyl is flat, cut away excess flooring. Run a utility knife along a straightedge to cut the vinyl where it meets the walls. Baseboard or shoe molding will then be installed to hide the joint between the floor and wall.

CARPET

Carpet is not a common floor covering for kitchens and bathrooms, but it is sometimes used. If after careful consideration you decide to install carpeting in your bathroom, make sure the pile of the carpet faces the entrance. This will enhance the appearance of your room. Most carpet is available in widths up to 12 feet.

If you want to install carpeting in your bathroom, weigh the pros and cons. While carpeting makes the floor warmer and less slippery than vinyl or tile, it retains moisture, and that can create problems with mold, mildew, and rot.

The installation of loop-pile and cut-pile carpeting can be done by homeowners, but you will probably have to rent a few tools. You will also have to be careful in your measurements and cuts.

CERAMIC TILE

Ceramic tile is often found in bathrooms. If you decide to use tile, you will have plenty of choices to choose from. Tile floors can be made of quarry tile, mosaic tile, and glazed ceramic tile. Quarry tile comes in large squares, and it is produced in natural clay colors. Mosaic tile is small and generally comes with numerous tiles connected to a single backing. Glazed ceramic tile may be bought as squares or rectangles.

Underlayment should be installed over a subfloor before the tile is installed. The underlayment should be at least 3/8-inch thick and should be installed with 1/8-inch expansion gaps between the sheets. The tile can be secured with adhesives. The adhesive may be organic or epoxy. Epoxy is the preferred adhesive for floors where dampness is a problem.

Grout material is a product that fills the gaps between tiles, preventing water and dirt from collecting in the voids. There are numerous types of grout material. Check with your tile dealer before selecting a grout material.

When installing tile, the choice of which type of adhesive to use is often determined by the manufacturer's recommendations. Check with your tile dealer for specifics. Always follow the manufacturer's recommendations.

Proper planning is a critical element of good tile installations. Deciding how to obtain the proper pattern and spacing will require thought. Special saws and cutters should be used to cut your tiles. These saws and cutters can be rented at tool centers.

Figure 13.3 Examples of ceramic tile. *Courtesy of Armstrong*

Installation methods vary. You should check with your dealer and follow the manufacturer's recommendations for the installation of your tile. Let's look at one common way of installing tile.

Trowel adhesive on the underlayment to a thickness of about ¼ inch. Use plastic spacers (available at tile dealers) to maintain even spacing between your tiles. Lay your first tile in the center of the floor and lay subsequent tiles out from that point.

As you set the tiles in the adhesive, press them down firmly. If needed, a rubber hammer can be used to tap the tile into place. Use a long level to check the consistency of the

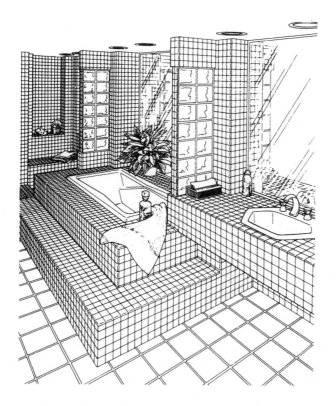

Figure 13.4 Excessive use of tile can be overpowering. *Courtesy of Dal-Tile*

floor; the tile should be installed level. After the tile is set, you must wait for the adhesive to dry.

When the adhesive is dry, you are ready to grout the tile. The grouting should be spread over the floor, filling all gaps between the tiles. This is usually done with a special trowel. Once the grout has filled the cracks, wash the remainder of the grouting off with a wet sponge. To be sure you install your tile properly, follow the tile manufacturer's recommendations.

Now that you know how to install the most popular types of flooring for bathrooms, let's move on to the next chapter and learn about walls and ceilings.

14

Walls and Ceilings

Walls and ceilings can account for much of the work required in remodeling a room. If finished walls are not removed, the work may only entail painting. But major remodeling often involves building new walls and recovering old walls. Under these conditions the effort required is considerable, and skills in many trades are needed.

You may be faced with framing new interior partitions or installing a new door or window. Insulation may need to be added to the exterior walls, and getting a good finish on drywall is something of an art. Even painting is not always as easy as it looks. This chapter is going to introduce you to the work that may be involved with your walls and ceilings.

FRAMING

Framing work is often called rough carpentry, but that doesn't mean you can use rough estimates or get by with rough skills. If the framing is not done properly, the rest of the job will suffer. For example, if you frame a new wall to hang cabinets on and the wall is out of plumb, you are going to have a tough time hanging the cabinets in a satisfactory manner. Some bathroom remodeling jobs don't require any framing work, but others do. Framing interior partitions is not difficult, but there are a few tricks of the trade that makes the job easier; let's see what they are.

Building a Wall

When you are building a wall, you can get the job done in several ways. Most professionals build walls by laying the framing on the subfloor and then standing the walls up. This process allows you to frame the entire wall under comfortable circumstances,

Before you start driving nails, lay out the wall locations on the subfloor. Mark the wall location with a chalk line. Once the wall location is known, measure the length of the proposed wall. This will tell you how long your top and bottom plates should be.

Carpenters normally use one 2-x-4-inch stud as a bottom plate and two studs for the top plate of a wall. Begin by cutting the bottom and top plates to the desired length. Next, measure to determine the length needed for vertical studs. Remember to allow for the thickness of your top and bottom plates when measuring between the ceiling joists and the sub-floor. After you are sure of your measurements, cut the wall studs to the desired length.

Turn the bottom and top plates over on their edges. Place the first wall stud at one end of the plates and nail it into place. Do the same with a second stud at the other end. You have created a rectangle, and all you have to do is install the additional wall studs. Studs are normally installed so that there are 16 inches from the center of one stud to the center of another. When the wall section is complete, you can stand it up and nail it into place.

When building walls that will meet at corners, you have to have a way to tie the two wall sections together. This is normally done with studs that are turned flat in the wall section. This provides a solid surface to attach to. You can screw or nail one wall section to the other in this manner.

Figure 14.1 A variety of wall treatments create a stunning effect. *Courtesy of Moen, Inc.*

Figure 14.2 A simple tile shower wall. *Courtesy of Moen, Inc.*

WINDOWS AND DOORS

Your framing work may involve windows and doors. It is not uncommon for new windows and doors to be installed during large remodeling jobs. Let's take a look at the types of windows and doors you might want to use and how to frame for them. Here are some of your options:

- Casement windows are well known for their energy-efficient qualities. This style of window offers the advantage of full airflow. When you crank out a casement window, the entire window opens.

- Double-hung windows are the type of windows found in most homes. These windows are generally less expensive than casement windows, and they are well accepted as an industry standard.

- Skylights can give a bathroom plenty of natural light. Rooms filled with sunshine generally appear larger and more inviting. Today's skylights are available with built-in shades and screens.

- Metal doors are relatively inexpensive, and they can be equipped with good insulation qualities. These doors are available as solid doors, stamped doors that give the appearance of a six-panel door, and half glass doors, with or without grids. These doors can be painted, but they cannot be stained.

- Wood doors typically cost more than metal doors, but they are available in more styles. Some people don't like wood doors because they can warp and become hard to operate.

- French doors are a wonderful way to brighten up an eat-in kitchen. These doors are beautiful, but they tend to be expensive.

FRAMING WINDOW AND DOOR OPENINGS

Framing window and door openings is simple when building new walls, but it can be complicated if you are cutting new windows or doors into existing walls. Your work will affect the siding on your home and the structural integrity of the exterior wall. The basics of framing window and door openings are about to be explained, but understand that existing conditions at your home may call for professional help in this phase of the project.

A typical window frame will involve jack studs, cripple studs, and a header. The header will provide strength and support for whatever is sitting on top of the wall. It is usually made with lumber that is nailed together. Jack studs are installed under the header to support it. A horizontal board is installed below the header at a distance equal to the rough-opening dimension for the window being installed. This board is nailed to the wall studs and supported with short studs from below. The area above the header is filled with cripple studs. These cripples extend from the header to the top plate, completing the window frame.

> Don't install headers that are too small. The headers carry weight and must be large enough to get the job done. It may be tempting to use leftover wood. Don't do it. Build your headers out of wood with the proper dimensions for the application.

> Framing for an exterior door is similar to framing for windows. Most doors are available as pre-hung units. These units come to the job ready for installation. It makes sense to buy pre-hung door units. The time saved offsets the additional cost.

The rough door opening should extend all the way to the subfloor. A header, jack studs, and upper cripples will be installed in a manner very similar to window framing. However, the lower framing that is done with a window is eliminated, and the section of the bottom wall plate that runs through the door opening is cut out. Cripple studs may not be used when a large header is installed.

WINDOW INSTALLATION

If you have framed your rough opening properly, window installation is simple. Many windows have nailing flanges for attaching the window unit to the frame walls. Sit the window unit in the rough opening and make sure it is plumb. Nail the unit in place by driving nails through the flange. The flange should be on the exterior side of the house. If you are working with windows that don't have flanges, you will be nailing through the window framing.

DOOR INSTALLATION

When you are working with a pre-hung door unit, installation is not too difficult. Put the unit in place and level it. It may be necessary to install shims around the frame to get it plumb. When the door is plumb, nail the jamb to the framed opening. Regardless of what you are installing, always read and follows the recommendations from the manufacturer of the product.

INSULATION

Insulation is not difficult to install, but it can irritate your skin. The easiest type to install for most remodeling jobs is batt insulation. This type of insulation is available in widths made to fit standard wall and joist cavities.

Wall insulation should have a vapor barrier. The barrier should be installed so that it is between the heated room and the insulation. You can buy rolls of batt insulation with a vapor barrier already attached, or you can use unfaced insulation and install sheets of plastic as a vapor barrier. It is important to have the barrier facing the heated room, not the outside of the house. A strong stapler is the only tool needed for installing insulation.

> Never install the vapor barrier for insulation towards the outdoors. This can cause condensation and rotting of the insulation and wood members in a wall section.

DRYWALL

Other than dealing with the weight of the material, hanging drywall is not difficult. Finishing it, however, does take some time and practice. When installing new drywall in a bathroom, moisture-resistant products should be used. Let's take a look at what is involved with hanging and finishing drywall.

Drywall is available in different sizes. Professionals often use 4-x-12-foot sheets to reduce the number of seams in a job, but 4-x-8-foot sheets are much easier for the average person to handle. You can choose from different thicknesses to give your finished wall the proper depth.

Drywall can be hung with the use of nails or screws. Screws are less likely to work loose than nails. If screws are used, an electric screwdriver makes the job go much faster. Screws should be tight enough to make a depression in the wallboard. When nails are used, they should be driven extra deep to create a dimple in the drywall. The depressions will be filled with joint compound to hide the nail and screw heads.

Drywall can be cut with a drywall saw, jigsaw, or utility knife. Most pros use utility knives. The procedure requires the drywall to be scored with the utility knife and then broken at

Figure 14.3   Simple walls provide the backdrop for a dramatic bathroom décor. *Courtesy of Moen, Inc.*

the scored seam. You can use a T-square, piece of lumber, or chalk line to make straight cuts.

Hanging drywall on a ceiling is the most difficult part physically of any drywall job. Ceilings should be hung before the walls are covered with drywall. When you drywall a ceiling, you will have to make cutouts for ceiling-mounted electrical boxes. Due to its weight, drywall is not easy to install above your head. However, there is a way to reduce this burden.

> It's possible to hang a ceiling by yourself. Sit the drywall on the tops of two ladders. Leave a couple of feet of the wallboard hanging over each end of the ladders. Put one T-brace under one end of the drywall and raise it with the brace. Wedge the brace against the floor and raise the other end of the drywall with another brace. This will take some time and practice, but once you get the hang of it, you can install your ceiling without help.

A T-brace will be of much assistance when hanging drywall on a ceiling. You can make a T-brace from scrap studs. To make the brace, nail a 2-x-4-foot piece of wood (about 3 feet long) onto the end of another 2 x 4 that is long enough to reach the ceiling, with a little left over.

The brace can be wedged under the drywall to hold it to the ceiling. The T-arm will rest under the drywall, and the long section of the brace will be wedged between the subfloor and the ceiling. It normally takes two people to raise drywall to the ceiling joists. Once the T-brace is wedged into place, it frees one of the people up to attach the drywall to the joists. Two braces can be used to free all hands for other work.

Hanging drywall on walls is much easier than on ceilings. The drywall can be hung vertically or horizontally. If you hang your walls vertically, you shouldn't need any help. Hanging the drywall horizontally generally results in fewer seams, but it is more difficult to do without a helper. There are, however, some tricks that make horizontal hanging easier for the sole remodeler.

Nail large nails to the studs to provide temporary support for the drywall panel. You can then rest the sheet of drywall on the nails while you attach it to the studs. The large nails can be removed once the drywall is secured.

A ledger can be used in place of nails for more uniform support of the drywall. Nail a 2 x 4 horizontally across the wall studs. Rest the drywall on the ledger while you attach it to the studs.

Outside corners of walls covered with drywall should be fitted with metal corner bead. The metal strips protect the exposed corners and edges. These strips are perforated and can be nailed or screwed to wall studs. The corner bead is designed to retain joint compound for a smooth finish.

Inside corners do not require metal corner bead. Drywall tape should be creased and installed to cover the seams of inside corners.

Taping the seams of new drywall is not difficult, but it may take a while to develop a feel for what you are doing. Buy joint compound that is premixed. The tape you will use to cover the seams does not have an adhesive backing; it is held in place by the joint compound. A wide putty knife (about 4 inches wide) should be used to spread the joint compound over the tape, seams, and dimples.

The first coat of joint compound should be spread over seams, corner bead, and dimples. It is best to work one seam at a time. The first layer of compound should be about 3 inches wide, and it should be applied generously.

Once the compound covers a seam, lay a strip of tape on the compound and use a putty knife to work the tape down into the joint compound. The tape should sit deeply into the compound. Smooth out the compound and feather it away at the edges of the tape. Continue this process on all seams.

Tape is not necessary when filling nail dimples or covering corner bead. Simply apply joint compound in the depressions until it is flush with the drywall. Smooth the compound out with your putty knife and let it dry.

The first layer of joint compound should dry within 24 hours. A second layer will be applied on top of the first layer; it should be about twice as wide. The second layer must be left to dry for about 24 hours.

A third layer of compound is usually the final finish on drywall. Before applying the last coat of compound, you must sand the second layer you installed.

Sand the compound first with medium-grit and then with fine-grit sandpaper. A good dust mask is very helpful during this job. A sanding block will make the job go faster and will be easier on your hands. Sand the compound with soft strokes to avoid scarring the walls.

> When hanging drywall on the wall studs, don't forget to leave cutouts for electrical boxes, water supplies, drain arms, and other items that should not be covered up.

When the second layer has been sanded properly, you may apply the third layer of compound. This last layer should be about 10 inches wide, and the edges should be feathered out. This final layer should be applied in a thin coat.

After the final layer has dried, it must be sanded. Use fine-grit sandpaper for the finish sanding. When this step is complete, you are ready to clean up and prepare to prime and paint the walls.

PAINT

Before you begin to paint, vacuum the room to remove all dust. If you don't, your paint will catch the dust, and the job will not look good. You will be working over a subfloor, so drop cloths are not necessary.

You must decide whether to use latex or oil paint. Latex cleans up better than oil, and it will do a fine job on your walls and ceiling. New walls and ceilings should receive at least one coat of primer and one coat of paint. When buying your primer, ask the paint dealer to tint it to match the finish color.

Most painters begin their work on the ceilings of a home. Paint rollers work well for applying paint and primer to ceilings. When you paint a ceiling with a roller, you have to cut in along the joints between the walls and ceiling with a brush. Use a 2- or 3-inch brush to apply a strip of paint to the edges of the ceiling. As soon as the strip of wet paint is applied, lay

Figure 14.4 Note the unusual ceiling in this bathroom. *Courtesy of Armstrong*

down the brush and pick up a roller. Use an extension handle on the roller to avoid numerous trips up and down a ladder.

With the ceiling finished, you are ready to paint the walls. Apply cut-in strips of fresh paint around the tops of the walls. Follow the same procedures used on the ceiling to avoid mismatched paint.

After the first coat of paint or primer you may see imperfections that had been invisible. Take time between the first and second coat to touch up the drywall. Vacuum any dust created from the touch-up work before applying the second coat of paint.

> New walls and ceilings should receive at least one coat of primer and one coat of paint. When buying your primer, ask the paint dealer to tint it to match the finish color.

You may wish to texture your ceiling. If so, there are many options available to you. Joint compound, just like that used to finish drywall, can be used to create a textured ceiling. Some types of paint are already texturized.

A variety of devices can be used to texture a ceiling. A stiff paintbrush can be used to create texture, and a stipple paint roller will also get the job done. Trowels can be used, and even common potato mashers are sometimes used to texture ceilings.

Roll paint on the ceiling and over the strip of fresh paint. Do the cut-in work a little at a time. Trying to cut in the whole ceiling before rolling on the paint will result in a mismatched finish. The cut-in strips will dry before the rest of the paint does. This results in two different finishes and looks strange. Roll paint on the ceiling in generous amounts; otherwise it will dry without covering the surface.

The trim work around your windows, doors, and walls will also need to be painted or stained. Paint with a gloss finish is often used on trim when paint with a flat finish is used on walls. Bathroom walls are often painted with gloss paint; it is easier to clean than flat paint.

Before the final paint or stain can be applied to trim work, nail holes must be filled in with putty. Wood putty and a small putty knife are all that is needed for this job. If you will be staining the trim, use putty that will not show through the stain.

Trim is often stained or painted before it is installed. If you are going to stain your trim, be sure to get clear wood for the trim material. Trim can be stained with either a staining mitt or a brush. After the trim has been stained, you may wish to apply sealer over the stain. This is not a required step, but some people prefer the look and durability offered by sealants. Walls and ceilings are not particularly difficult to work with. However, they are important elements in a job and should be given the respect that they deserve.

Once your drywall is finished and ready for paint, prime it. One coat of primer will expose flaws in the finish work that you may not be able to see before the primer is applied.

15

Mechanical Work

S ome types of mechanical modifications are generally required in all bathroom remodeling jobs. Not all renovations and alterations are big jobs, but even minor modifications can present major problems for inexperienced people.

A plumbing connection that is not made properly may blow apart, flooding the room being remodeled and the rest of the house. A mix-up in electrical wiring can be difficult to locate and can cause a number of problems. Moving a floor register to gain better heating can result in serious cuts from the sharp metal in ductwork.

Little jobs can create big problems. Most of the problems can be avoided and many of them are easy to correct, but you must have the right knowledge. Knowing this, let's look at what you may be involved with when working on your plumbing, heating, and electrical systems.

PLUMBING PIPES USED FOR DRAINS AND VENTS

There are many types of plumbing pipes that may be used for drains and vents. Most modern plumbing is done with plastic pipe, but older homes may be plumbed with pipes made from the following materials:

- Cast iron

- Galvanized steel

- Brass

- Lead

Some of these pipes do not perform well once they age. If you open up your walls during remodeling, it may pay to replace sections of your plumbing to avoid future problems. Let's look at the most common types of pipe found in remodeling jobs to see their strong points and weaknesses.

Cast-Iron Pipe

Cast-iron pipe can be found in houses of all ages. It has not been used in residential plumbing extensively since the mid-70s, but cast iron is still used today. If your home is more than thirty years old, there is a good chance it may have cast-iron drains and vents.

Cast iron was typically used for large drains and vents. Galvanized-steel pipe was often used in conjunction with cast iron for small drains, such as those in kitchen sinks, bathtubs, and lavatories. You will most likely work with galvanized-steel pipe in simple remodeling jobs.

If you are forced to cut cast-iron pipe, you will be well served to rent a cutting tool for the job. It's possible to cut cast iron with a chisel or hacksaw, but a chain-type cutter is far easier and faster to work with. And, the cuts are cleaner.

Cast-iron joints used to be made with oakum and molten lead, and they still are today. However, technology has delivered special rubber adapters for making connections with cast iron in modern installations. There are three basic types of rubber adapters. One type resembles a doughnut and is placed in the hub of one pipe so that the end of another pipe can be inserted, making a watertight joint. The other two types are used with cast-iron pipe that does not have hubs. These adapters slide over the ends of two pipes and are held in place with stainless-steel clamps. Not only is this type of connection much easier to make; it is also safer than working with hot lead.

Unless you are relocating a toilet or altering the main drainage and vent system in your home, it is unlikely that you will have to work with cast-iron pipe. But since you may wish to relocate a toilet or do some other type of plumbing remodeling, let's take a quick look at how you can simplify the task of working with cast-iron pipe.

If you plan to cut cast-iron pipe, rent a ratchet-type soil-pipe cutter. This tool makes quick, easy work of cutting cast iron. All you have to do is wrap a special cutting chain around the pipe, secure the cutter, pump the handle a few times, and the pipe is cut cleanly. This is much easier than laboring your way through the pipe with a hacksaw.

> Cutting vertical sections of cast-iron pipe can be dangerous. If the pipe is not supported properly, the vertical piping could come crashing down on you. Before you cut a vertical pipe, make sure it is supported in a way to protect you.

Galvanized-Steel Pipe

If you have cast-iron pipe in you home, you probably also have some galvanized-steel pipe, too. This pipe tends to rust and build up blockages over the years. If you have the opportunity to replace galvanized pipe with plastic pipe, do it. You will be saving yourself from future trouble.

> If you want to convert a piece of cast-iron pipe to another type of pipe such as plastic pipe, use a universal rubber adapter for the conversion. This will make the job fast and simple.

Galvanized pipe can be cut with a hacksaw, and the same rubber adapters used to join cast-iron and plastic pipe can be used on galvanized pipe.

DWV Copper

DWV copper was a popular drain and vent pipe for many years, and it is still found in many older homes. On the whole, copper drains and vents give very good service and should not need to be replaced. Copper drains and vents can be cut with a hacksaw, and the same universal adapters used with cast-iron and galvanized pipe can be used to convert copper to plastic.

Schedule-40 Plastic Pipe

Schedule-40 plastic pipe is the drain and vent pipe most often used in modern plumbing systems. There are two types of schedule-40 plastic pipe used in homes: ABS and PVC. ABS is black and PVC is white.

Both these pipes are easy to work with, and either can be cut with a hacksaw or a standard carpenter's saw. Joints for these pipes are normally made with a solvent or glue. Most plumbing codes recommend that a cleaner be used on plastic pipe and require that a primer be applied prior to gluing a joint. These pipes can be joined to any of the other types of drains and vents mentioned with universal rubber adapters.

PLUMBING PIPES FOR POTABLE WATER

Just as there are a number of approved drain and vent materials, there are also several types of plumbing pipes for potable (drinking) water. Let's take a quick look at some of them.

Copper

Copper water pipe and tubing are found in more homes than any other type of water-distribution pipe. Copper is a dependable material that provides years of service. It can be cut with a hacksaw, but roller-cutters will cut the pipe much smoother. The joints for copper pipe and tubing are usually made by soldering. This can be a problem for some homeowners. Learning to solder watertight joints takes some time and experience. One way to avoid soldering is to use compression fittings.

Compression fittings are available in all shapes and sizes. They are easy to install, and they normally don't leak. If your joints are going to be concealed in a wall, compression fittings may not be a good idea, but they work well under sinks and in other accessible areas.

> There is some risk that compression fittings will develop leaks as the pipes are vibrated with use. The leaks will be small and can be seen and corrected easily if they are visible. If, however, they are concealed in a wall, a small leak could go on for a long time, causing serious damage to building components before it is detected.

CPVC Pipe

CPVC pipe is another alternative for homeowners lacking soldering skills. CPVC is a rigid plastic pipe that is put together with solvent joints. A cleaner and primer should be used on the pipe and fittings prior to gluing joints.

CPVC can be cut quite easily with a hacksaw, and it is simple to install. You should, however, allow plenty of time for joints to dry before moving the pipe. If a fresh joint is bumped or twisted before the glue has dried, a leak is likely.

> When working with CPVC in cold temperatures, be careful not to drop the pipe on hard surfaces such as cement floors. CPVC tends to be very brittle under cold conditions. The pipe can crack if dropped. The crack may not be visible until the pipe is pressurized with water.

PEX Pipe

Pex pipe is the new kid on the block. It is a flexible plastic pipe than can be installed much like electrical wiring. The pipe can be snaked through studs, and its flexibility allows for minimum joints.

If you opt for Pex pipe, you will need to rent a special crimping tool. Do not attempt to make joints with standard stainless-steel clamps. Pex joints require the use of insert fittings and special crimp rings. Many professional plumbers feel that Pex will eventually be more common than copper for potable water systems.

WHERE SHOULD YOU PUT YOUR PIPES?

Where should you put your pipes? The locations for pipes will vary with the type of fixtures being plumbed. Your plumbing supplier should be able to provide you with a rough-in book. The rough-in book will tell you exactly where to place your pipes. Exact rough-in measurements are usually not critical, but they can be, especially with fixtures such as pedestal lavatories. While it is impossible to predict exactly where your pipes should go without rough-in specifications, there are some rules-of-thumb that will normally work. Let's take a fixture-by-fixture look at where you might want to put your pipes.

Lavatories

Drains for lavatories should come out of the wall about 17 inches above the subfloor. The center of the drain should line up with the center of the lavatory. For example, if you were installing a vanity against a sidewall and the center of the lavatory were 15 inches away from the wall, the drain should also be about 15 inches away.

If the drain is roughed in too low or too far to the left or right, the problem can be corrected. A tailpiece extension will compensate for a low trap, and fittings can be used to bring the trap arm closer to the fixture trap. However, if the drain is roughed in too high, you've got a problem that is not so easily corrected. The only solution to this problem is to rework the rough plumbing completely.

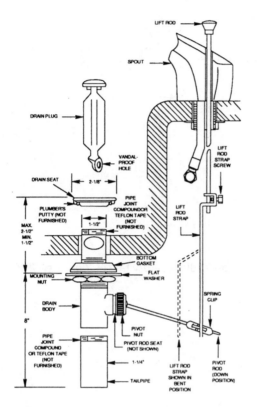

Figure 15.1 Detail of a pop-up assembly for a lavatory. *Courtesy of Moen, Inc.*

Water pipes for lavatories, if they come out of a wall, should be about 21 inches above the subfloor. Water pipes that come up out of the floor can be extended as needed. Hot-water pipes should always be installed on the left side (as you face the lavatory to use it).

Most lavatories have faucets with 4-inch centers. This simply means that there are 4 inches between the hot and cold water when measured from the center of one supply tube to the other. If the drain is roughed in under the center of the lavatory, each water pipe will be about 2 inches from the center of the drain on each side.

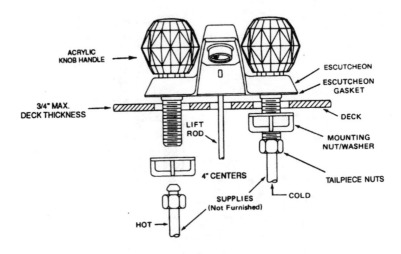

Figure 15.2   Two-handle lavatory faucet. *Courtesy of Moen, Inc.*

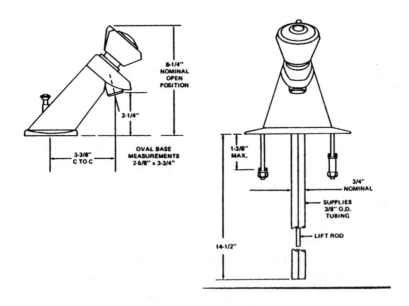

Figure 15.3   Single-handle lavatory faucet. *Courtesy of Moen, Inc.*

Toilets

Unless you are relocating your toilet, you will have no need to rough in a new drain. However, if you are installing a new closet flange (the part a toilet sits on), the center of the flange should measure 12½ inches from the back wall behind the toilet. This measurement is based on measuring from a stud wall. If you are measuring from a finished wall, the distance would be an even 12 inches.

> When buying lavatory faucets, make sure that the faucets you plan to buy will fit your lavatory. Not all lavatories accept the same size faucets. Most lavatories require a faucet with a 4-inch center, but some use 8-inch-center faucets.

You should be able to measure 15 inches from the center of the closet flange to either side without hitting a wall or other fixture. Toilets are required to have a minimum free width of 30 inches for proper installation.

Water supplies for toilets should be installed 6 inches above the subfloor, and they should be 6 inches to the left of the center of the drain.

Bathtubs

Drains for bathtubs are typically located 15 inches off the stud wall where the back of the tub will rest. The drain is normally about 4 inches off the head wall where the faucets will go. If you are installing a new tub drain, it is necessary to cut a hole in the subfloor for the drain before the tub is set in place permanently. The hole should extend from the head wall to a point about 12 inches away, and it should be about 8 inches wide. This gives you a hole that is 8 inches wide and 12 inches long. You will need most of this space to connect a tub waste and overflow.

Bathtub faucets should be installed about 12 inches above the flood-level rim of the tub. The flood-level rim is the armrest or location where water will first spill over the tub. Tub spouts are normally mounted about 4 inches above the flood-level rim of the tub.

When installing the faucet for a tub-shower combination, the shower head outlet should be set about $6\frac{1}{2}$ feet above the subfloor.

Showers

A shower should not be installed permanently until a hole has been cut in the subfloor for the shower drain. Most shower drains are located in the center of the shower, but there are many types of showers where this is not the case. Consult a rough-in book or measure the actual drain location to determine where to rough in the shower drain.

Faucets for showers should be installed about 4 feet above the subfloor. The shower head outlet should be installed about $6\frac{1}{2}$ feet above the subfloor.

HEATING SYSTEM MODIFICATIONS

Heating system modifications are not needed in most bathroom remodeling jobs. Unless you are expanding the size of your room, the existing heat should be adequate and require no major work. However, there are times when heat needs to be relocated within the room. Assuming that there is good access from under the floor, moving heat around is not normally a big job.

Ductwork used in forced-hot-air heating and central air conditioning has very sharp edges when disassembled. If you are going to work with heating or air-conditioning ducts, wear gloves.

Ductwork is usually held together by metal strips that slide into a channel. These strips can be dislodged with a hammer. Metal-fabrication shops will be glad to make lengths of ductwork or offsets to your specifications. Other than the risk of cutting yourself on sharp metal, installing ductwork is not too difficult.

If you will be adding new ducts to an existing system or extending the length of existing ducts, talk with some profes-

Rough-In Measurements

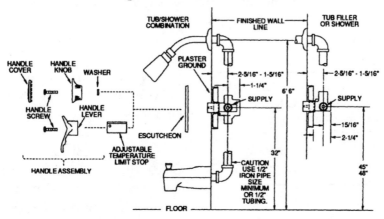

Figure 15.4   Rough-in measurements for tub-shower faucet. *Courtesy of Moen, Inc.*

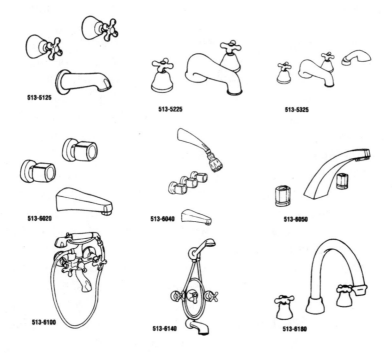

Figure 15.5   Tub and tub-shower faucets. *Courtesy of Eljer*

sionals beforehand. Typically, the size of ductwork gets smaller as it goes along its route. If the duct is not sized properly, it cannot perform to its optimum output. These types of alterations may affect the effectiveness of your heating and cooling system. New ducts can be cut into existing trunk lines easily, but you must be sure your alterations will not strain your system.

In some cases flexible ducts can be used to carry heat or cool air from a main trunk to an outlet register. Flexible duct is obviously easier to work with, and you are not as likely to hurt yourself.

Hot-water heat, in modern homes, runs through copper tubing similar if not identical to the type used for potable water distribution. Don't attempt to work with this type of heat unless you know how to solder. Before you cut into heating pipes, make sure the boiler is turned off and drained down to a point below the location on the pipes where you will be cutting.

There are two basic types of hot-water heating system: one-pipe systems and two-pipe systems. In a one-pipe system, a supply pipe leaves the boiler and runs to the first heating unit. The water passes through the heating unit; when it comes out, it is conveyed to the next heating unit through more supply pipe. Some people refer to this as a loop system because the supply pipe makes a big loop through all the heating units and back to the boiler.

Two-pipe systems rely on supply pipes and return pipes. These systems are more costly, due to the extra pipe involved, but they produce better heat. In these systems each heating unit receives a supply of hot water from one pipe, and another pipe immediately returns the water directly to the boiler.

Adding new heat to an expanded bathroom is certainly possible, but you must make sure the boiler is capable of heating the extra space. It would be very unlikely that any existing boiler couldn't handle the small amount of space being added for a bathroom, but you should make sure before altering the heating system.

After the new hot-water heating units are installed, you will need to bleed air out of the system. Professionals usually install special elbows, called vent ells or bleed ells, at individual heating units. These fittings are most commonly installed on the heat that has the highest elevation in the home. If your home has more than one story, bleed air from the highest heating units.

All you have to do in order to bleed air from the heating system is to cut the boiler on and remove the cap from a bleed fitting. You will probably hear air hissing out of the opening in the fitting. When the air is replaced with a stream of water, you have removed the air from the system.

It is customary with all types of heat to install it on exterior walls, usually under windows. If your house has an old heating system that works with steam or radiators, you should probably call in professionals to make the necessary changes in the system. These systems can be troublesome to work with, and it is not unusual for radiators and old steel pipes to fail and leak when disturbed.

ELECTRICAL SERVICES

Sometimes homeowners either want to or have to upgrade their electrical services for major remodeling projects. This is not normally the case with bathroom remodeling, but it can be. This is not a job you should do yourself unless you are a well-trained electrician with experience in working with panel boxes. All work with electricity poses some danger, but the risks involved with replacing a service panel are too great to take.

Installing New Electrical Boxes

Homeowners who are competent to work with electricity can install their own electrical boxes, but if you do not know what you are doing, leave all electrical work to licensed professionals. Poor workmanship with electrical wiring can result in fatal shocks and houses being burned to the ground. If you will be working with electrical boxes, here are some types and their uses:

- Switch boxes are usually rectangular. These boxes are commonly used for wall outlets and wall-mounted lights. The dimensions for rectangular boxes are generally 3 inches by 2 inches.

- Boxes used for ceiling lights are often octagonal. These boxes may also be used as junction boxes for joining numerous wires together. Each side of the boxes is typically 4 inches long. Round boxes are also used for ceiling lights.

- Both octagonal and square boxes are used as junction boxes. Square boxes are more common and have typical dimensions of about 4 inches.

- Depth requirements for electrical boxes are determined by the number of wires to be placed in the box. Common depths vary from just over 1 inch to about $3\frac{1}{2}$ inches.

Mounting Electrical Boxes

The mounting of electrical boxes can be done in a number of ways. Some boxes are sold with nails already inserted. All you have to do is position the box and drive the nail into a piece of wood. Other boxes have flanges that nails are driven through. Some boxes have flanges that move and allow more flexibility in terms of installation.

If you are going to install your own electrical boxes, you must decide which types of boxes to use. The size and shape of electrical boxes vary with their purpose. Choose the proper type of box for the type of work that you will be doing with it.

Boxes for ceiling fixtures are often nailed directly to ceiling joists. If the boxes need to be offset, such as in the middle of a joist bay, metal bars can be used to support the boxes. The metal bars are adjustable and mount between ceiling joists or studs. Once the bar is in place, the box can be mounted to the bar.

Rough-In Dimensions

Rough-in dimensions can be determined by local code requirements and the fixtures to be served. There are, however, some common rough-in figures you may be interested in knowing about.

Wall switches are usually mounted about 4 feet above the finished floor. Outlets are normally set between 12 and 18 inches above the floor and are spaced so that there is not more than 12 feet between outlets.

Where Does the Red Wire Go?

Where does the red wire go? What should be done with the black wire? These are questions many people have about wiring. Electrical wires are insulated with different colors for a purpose. The colors indicate what the wire is used for and where it should be attached.

Black and red wires are usually hot wires. White wires should be neutral, but they are sometimes used as hot wires. Don't trust any wire not to be hot. Green wires and plain copper wires are typically ground wires.

When matching colored wires to the screws in an electrical connection, they should be installed something like this. Black wires should connect to brass screws. Red wires should connect to brass or chrome screws. White wires are normally connected to chrome screws. Green wires and plain copper wires should connect to green screws.

Electrical wires should be crooked and placed under their respective screws in such a way that the crook in the wire will tighten with the screw. In other words, the end of the crooked wire should be facing in a clockwise position under the screw.

Wire nuts should be used when wires are twisted together. The colors of wire nuts indicate their size. Wire nuts are plastic on the outside and have wire springs on the inside. When wires are inserted into the wire nut, the nut can be turned clockwise to secure the wires. It is important to use a wire nut of the proper size, and it should be installed to a point where no exposed wiring is visible.

Ground Fault Interrupters

Ground fault interrupters (GFIs) are generally required in locations where a source of water is close to an electrical device. Bathrooms are required to be equipped with GFI circuits or outlets. GFIs are safety devices. They kill the power to an electrical device if moisture is detected. It is possible to install GFI outlets or GFI circuit breakers. Check with your local code-enforcement office to determine the requirements in your area.

You should now have a fair understanding of what may be involved with mechanical modifications in your home. Never attempt to do any type of work that you are not sure you can do safely. The price of professionals is a much better option than doing harm to yourself or your house.

16

Cabinets, Countertops, Fixtures, Trim, and Appliances

C abinets, countertops, fixtures, and trim must all be installed before your job will be finished. These aspects of the job play a vital role in how the finished job will look. They can also be more difficult to install than they appear to be. When you begin installing these items, you should be prepared to take your time. Carelessness in screw selection can cause you to puncture and ruin the surface of a new countertop or cabinet.

As you move toward the final phases of your job, you are likely to get excited by the thrill of completion. As you see all the cabinets installed, it can be tempting to work into the night to get the countertop installed. The desire to see the top installed before morning can be a mistake. If you are tired, you are more likely to make mistakes. The finish work in your project is very important; don't rush it.

Of the items discussed in this chapter, cabinets are normally the first to be installed. Cabinet installation requires attention to detail, but it is a job most handy homeowners can handle. Let's begin our look at the finish phases of work with cabinets.

Figure 16.1   A functional yet beautiful bathroom. *Courtesy of Wellborn Cabinet, Inc.*

CHOOSING CABINETS

Before you begin choosing cabinets, you owe it to yourself to shop around. You may be amazed at the number of variations available. You will have decisions to make on sizes, styles, colors, and features.

Cabinets are fundamental elements of many bathrooms. They are likely to attract as much attention as any other feature in the room. They will also receive a lot of use. Since you will want your bathroom to be beautiful, functional, and enjoyable to work in, take your time choosing the cabinets.

Cabinet materials can consist of solid wood, plywood, and particleboard. Many production cabinets use a mixture of these materials. Deciding on what you want your cabinets made of is only part of the buying decision. You will also have to look at the construction features of the cabinets. For example, dovetail joints should last longer than butt joints.

Custom cabinets are generally much more expensive than production cabinets. With the wide selection of production cabinets available, there is rarely a need for custom cabinets. Some people want their cabinets built just for them, but most people will have no trouble finding stock cabinets to suit their needs and desires. It is your decision.

Other considerations for choosing cabinets include whether the cabinet will have doors, drawers, or special accessories. An important consideration in choosing a drawer base is how well the drawers glide. Insist on a cabinet with quality glides and rollers.

An important consideration in choosing a drawer base is how well the drawers glide. Insist on a cabinet with quality glides and rollers.

CHOOSING WALL CABINETS

Choosing wall cabinets will be similar to choosing base cabinets. You will have to consider the sizes and styles that best suit your requirements. What will you want your cabinets to offer? The questions you'll need to answer include the following:

Look for quality in the shelves and latches of wall cabinets. The supports for shelves should be adjustable and allow random spacing. Magnetic latches are usually favored over plastic latches. Inspect hinges, structural supports, and all other aspects of wall cabinets before you buy them.

- Will your cabinets have glass doors?

- Do you want raised-panel doors?

- Will the cabinet doors have porcelain pulls?

- Do you prefer cabinet doors with finger grooves?

- There are plenty of choices to contemplate with cabinets.

CABINET INSTALLATION

Cabinet installation should begin with a design that you have studied and approved. It is much easier to make changes in a bathroom design on a drafting table or computer than it is in the bathroom. You should have a good cabinet layout already drawn. Any good cabinet supplier will provide recommended designs and drawings. When you are ready to install your cabinets, follow the design, but before you jump right into setting and hanging cabinets, double-check your previous work.

Check your floor and walls to make sure they are plumb and level. Cabinets that are not installed level may not operate properly, and there may be visual evidence of the poor installation. Shims can be used to overcome minor problems with walls and floors, but you should know what you are dealing with before you begin installing your cabinets.

Installing Cabinets

Install wall cabinets first. By installing the wall cabinets first, you reduce the risk of damaging base cabinets, and you will have more freedom of movement for the job. Before you put the cabinets in place, find the wall studs. If you have gutted your bathroom and installed new drywall, you will be familiar with the stud locations. You may have even thought ahead and marked their location on your cabinet plan for easy reference.

If you have trouble locating the studs, don't hesitate to probe the wall where the cabinets will be hung. The back of the cabinet will conceal any holes you make in the wall. A stud finder, which can cost less than $20, is also a major asset when seeking hidden wall studs.

When installing base cabinets, start with a corner cabinet, if there is one, and build out with the remaining cabinets. Base cabinets should be attached to each other in the same way as wall cabinets. Check frequently to see that the base cabinets are level and plumb. It may be necessary to shim under the cabinets to keep them level.

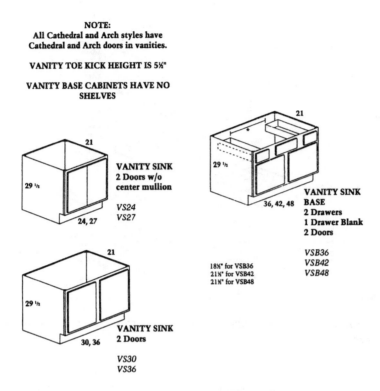

NOTE:
All Cathedral and Arch styles have
Cathedral and Arch doors in vanities.

VANITY TOE KICK HEIGHT IS 5½"

VANITY BASE CABINETS HAVE NO
SHELVES

Figure 16.2 Vanities with double doors and drawer blanks. *Courtesy of Wellborn Cabinet, Inc.*

COUNTERTOP INSTALLATION

Once you have your base cabinets set, you will want to install the countertop. Some people wait until the base cabinets are in to order their countertops. Working in this manner slows down the progression of the job, but it eliminates much of the risk of getting a countertop that is not sized properly. If you are buying your cabinets and countertop from a good supplier, your layout was probably drawn well in advance, and it is likely you already have the countertop. Assuming that you have your counter, let's see how it should be installed.

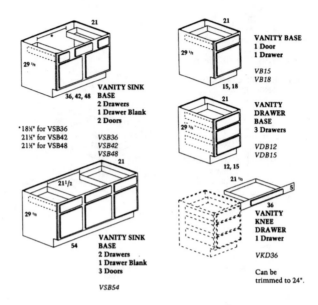

Figure 16.3   Vanity drawer bases and side cabinets. *Courtesy of Wellborn Cabinet, Inc.*

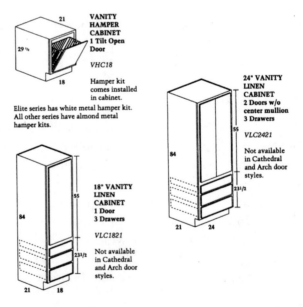

Figure 16.4   Vanity linen closets. *Courtesy of Wellborn Cabinet, Inc.*

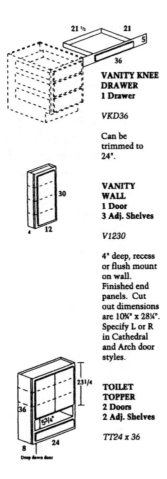

Figure 16.5 Accessory wall cabinets. *Courtesy of Wellborn Cabinet, Inc.*

Installing a Counter That Must Be Screwed Down

Look down on your base cabinets. You should see some triangular blocks of wood in the corners. These triangles provide a place to attach the counter to the cabinet. Before setting the counter in place, drill holes through these mounting blocks. Keep the holes in a location that will allow you to install screws from inside the cabinet. You may want to drill the holes on an angle towards the center of the cabinet. This will make the installation of screws easier.

Position the countertop on the base cabinets and check its fit. When you are satisfied with the positioning of the top, install screws from below. The screws used should be long enough to penetrate the triangular blocks and the bottom of the countertop, but be certain they are not long enough to come through the surface of the counter and ruin it.

Some counters must have a hole cut for the lavatory bowl. The supplier of the top will often cut this hole if he or she is provided with information on the size of the fixture. If you must cut your own hole, use the template that came with your new lavatory bowl. If you don't have a template to work with, turn the fixture upside down and sit it in place on the counter.

Lightly trace around the fixture rim with a pencil. Remove the bowl and draw a new outline inside the original tracing. The hole must be smaller than the lines you traced around the sink. There must be enough counter left after the hole is cut to support the rim of the lavatory.

When you are ready to cut out the hole, drill a hole in the countertop within the perimeter of the sinkhole. Use a jigsaw to cut out the hole. Put the blade in the hole you drilled and slowly cut the hole. Remember, the hole you make must be smaller than the outline of the sink. After the hole is cut, set the fixture in it and check the fit. You may have to enlarge the hole a little at a time to get a perfect fit.

INSTALLATION OF INTERIOR TRIM

The installation of interior trim is not difficult, but it does require precise measurements and patience. A miter box and back saw will be needed for cutting the angles required for interior trim. An electric miter saw is a better option. You can buy such saws for a few hundred dollars. Your decision depends upon whether you are willing to use elbow grease with a back saw or if you prefer an electrical saw. Once you get the hang of cutting angles, installing trim won't be much of a chore.

Baseboard trim should be nailed to wall studs with small finish nails. When baseboard trim meets a door casing or a cabinet, it simply butts against it. Shoe molding is generally installed with baseboard trim when vinyl flooring is used. Shoe molding is small trim that is installed in front of baseboard trim. It is often used to cover the joints where vinyl flooring meets a baseboard. If the flooring was installed before the baseboard trim, shoe molding is not necessary.

Windows, doors, and open entryways are often trimmed with casing. The only trick to installing this trim is in cutting the proper angles, and a miter box or miter saw will make that part of the job nearly foolproof.

If you plan to stain your trim material, do not buy finger-joint trim. The joints will show through the stain and look horrible!

The nails installed in trim should be countersunk. A nail punch can be used to drive the nail heads deep into the trim. Putty should then be placed in the nail holes before the final paint or stain work is done. If the trim is to be stained, make sure it is made of clear wood and that the putty will not show through the stain.

SETTING PLUMBING FIXTURES

When you are ready for setting plumbing fixtures, the end of your job is in sight. Some plumbing fixtures, such as toilets, must be handled with care, but the installation of most plumbing fixtures is not very difficult. Let's see what is involved with the installation of common plumbing fixtures.

Toilets

Toilets sit on and are bolted to closet flanges. Unless you have relocated the drain for your toilet, the existing closet flange should be able to be used. If you must install a new closet flange, install it so that the slots in the flange will allow the closet bolts to sit on either side of the center of the drain. The top of the flange should be flush with the finish flooring. If it is only slightly above the flooring, you should not have any problems, but it sits too high above the floor, the toilet will not mount properly.

Place closet bolts in the grooves of the flange and line them up with the center of the drain. Then install a wax ring over the drain opening in the flange. Now set the toilet bowl on the wax and press down firmly. The closet bolts should come up through the mounting holes in the base of the toilet.

Measure the distance from the back wall to the holes in the toilet where the seat will be installed. The two holes should be an equal distance from the back wall. If they are not, adjust the bowl until the holes are the same distance from the back wall.

Install the flat plastic caps that came with the toilet over the closet bolts. If metal washers were packed with the closet bolts, install them next. Install the closet-bolt nuts and tighten them carefully. Too much stress will break the bowl. Snap the plastic cover caps that came with the toilet over the bolts and onto the flat plastic disks you installed over the bolts. If the bolts are too long to allow the caps to seat, cut the bolts off with a hacksaw.

Uncrate the toilet tank and install the large sponge washer over the threaded piece that extends from the bottom of the tank. Then install the tank-to-bowl-bolts. To do this, slide the heavy black washers over the bolts until they reach the heads of the bolts. Push the bolts through the toilet tank.

Pick the tank up and set it in place on the bowl. The sponge gasket and bolts should line up with the holes in the bowl. With the tank in place, slide metal washers over the tank-to-bowl bolts from beneath the bowl. Follow the washers with nuts and tighten them. Again, be careful. Too much stress will crack the tank. Alternate between bolts as you tighten them. This allows the pressure to be applied evenly, reducing the chance of breakage. Tighten the bolts until the tank will not twist and turn on the bowl. With this work done, you are ready to connect the water supply.

Turn the water to the toilet's supply pipe off. Cut the supply pipe off about ¾ inch above the floor or past the wall, depending upon where the pipe is coming from. Slide an escutcheon over the pipe and install a cut-off valve. Compression valves require the least amount of skill and effort to install.

You will install a closet supply between the cut-off valve and the ballcock (the threads protruding past the bottom of the tank in the left front corner). Plastic supplies are the easiest type to install, but only nylon compression ferrules should be used with plastic supplies. Remove the ballcock nut from the threads at the bottom of the toilet tank. Hold a closet supply in place and, after measuring it, cut it to a suitable length.

Slide the ballcock nut onto the supply tube with the threads facing the toilet tank. Slide the small nut from the cut-off valve onto the supply tube and follow it with the compression ferrule. Hold the supply up to the ballcock and run the ballcock nut up hand tight. Insert the other end of the supply into the cut-off valve. Slide the ferrule and compression nut down to the threads on the cut-off and tighten the nut. Then tighten the large ballcock nut.

> When tightening closet bolts, tighten them evenly. Do not tighten one bolt down hard before tightening the other bolt; tighten them evenly. Otherwise you can break the tank.

Toilet seats generally have built-in bolts that fit through holes in the bowl. Put the seat in place and tighten the nuts that hold it in place.

This completes the toilet installation. After turning the water on, flush the toilet several times and check all connections to make sure none is leaking.

Lavatories

How you install your lavatory will depend on the type you are using. Drop-in lavatories require a hole to be cut in the counter where they will be mounted. The bowls are set in the hole and are held in place by their weight and plumbing connections. A bead of caulking should be placed around the hole, on the surface of the counter, before these lavatories are set in place.

Rimmed lavatories also require a hole to be cut in the countertop, but they are not as easy to install as drop-in lavatories. Rimmed lavatories have a metal rim that is placed in the hole of the countertop. The lavatory bowl is then held up to the ring from below and secured with special clips.

Wall-hung lavatories hang on wall brackets. Wood backing must be installed during the rough-in phase so that there will be a firm surface to bolt the wall bracket to. Once the wall bracket is mounted, most wall-hung lavatories just sit on the bracket. A few types have additional mounting holes where lag bolts can be used to provide additional security that the bowl will not be knocked off the wall bracket.

Vanity tops with the lavatory bowl built in are the easiest to install. These tops simply sit on a vanity cabinet. Normally the tops are heavy enough to sit in place without any special attachments being required.

Lavatory Faucets and Drains

All standard lavatory faucets and drains go together in about the same way. This job will be easier if you mount the faucet and drain assembly before you install the lavatory. A basin wrench may be necessary when working with faucets. It can be purchased inexpensively at any tool store.

Many faucets come with gaskets that fit between the base of the faucet and the lavatory. If your faucet doesn't have one of these gaskets, make a gasket from plumber's putty. Roll the putty into a long, round line and place it around the perimeter of the faucet base.

Place the faucet on the lavatory with the threaded fittings through the holes. Slide the ridged washers over the threaded fittings of the faucet and then screw on the mounting nuts. Tighten these nuts until the faucet is firmly seated.

Lavatory supply tubes mount on the ends of the threaded fittings protruding below the lavatory. Slide supply nuts up the supply tubes and screw them onto the threaded fittings. The beveled heads of lavatory supplies prevent leaks.

Now you are ready to connect the drainage. The first step is the assembly and installation of the pop-up drain assembly. Detailed instructions for the proper installation of the pop-up assembly should be packed with your faucet. Read and follow the instructions provided by the manufacturer.

Most wall-hung lavatories are made to accept legs, but the legs are optional.

A pop-up assembly mounts in the hole in the bottom of a lavatory. Unscrew the round trim piece from the shaft of the pop-up. This is the piece that you are accustomed to seeing when you look into a lavatory. Roll up some plumber's putty and place a ring of it around the bottom of the trim piece. Slide the fat, tapered black washer that is on the threaded portion of the assembly down on the threaded shaft. You may have to loosen the big nut that is on the threads to get the metal and rubber washers to move down on the assembly.

Apply pipe dope or a sealant tape to the threads of the pop-up assembly. With your hand under the lavatory bowl, push the threaded assembly up through the drainage hole. Screw the small trim piece, the one with the putty on it, onto the threads. Push the tapered gasket up to the bottom of the lavatory. Tighten the mounting nut until it pushes the metal washer up to the rubber washer and compresses it. You should notice putty being squeezed out from under the trim ring as you tighten the nut.

When the mounting nut is tight, the metal pop-up rod that extends from the assembly should be pointing to the rear of the lavatory bowl. Take the thin metal rod, the rod used to open and close the lavatory drain, and push it through the small hole in the center of the faucet.

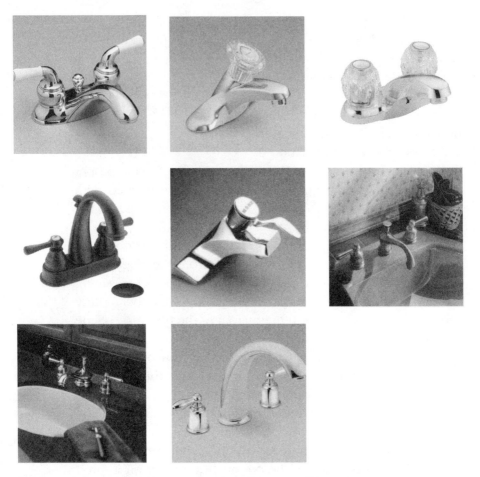

Figure 16.6   A variety of lavatory faucets. *Courtesy of Moen, Inc.*

You should see a thin metal clip on the end of the rod that extends from the pop-up assembly. Remove the first edge of this clip from the round rod. Take the perforated metal strip that was packed with the pop-up and slide it over the pop-up rod. You can use any of the holes for starters. Then slide the edge of the thin metal clip back onto the pop-up rod; this will hold the perforated strip in place.

At the other end of the perforated strip there will be a hole and a setscrew. Loosen the setscrew and slide the pop-up rod,

Figure 16.7 Assembling a faucet. *Courtesy of Moen, Inc.*

the rod used to open and close the drain, through the hole. Hold the rod so that about 1½ inches protrude above the top of the faucet. Tighten the setscrew. Pull up on the pop-up rod and see that it operates the pop-up plug. The pop-up plug is the stopper in the sink drain. You can test this best after all connections are made to the water and drainage systems.

There should be a 1½-inch chrome tailpiece (a round tubular piece) packed with the pop-up assembly. The tailpiece will have fine threads on one end and no threads on the other. Coat the threads with pipe dope or sealant tape. Screw the tailpiece into the bottom of the pop-up assembly.

Now you are ready to install the trap. First slide an escutcheon over the trap arm (the pipe coming out of the wall). Lavatory traps are normally 1½ inches, however, you can use a 1½-inch trap with a reducing nut on the end that connects to the tailpiece. Assuming that you used plastic pipe for your rough-in, you may either glue your trap directly to the trap arm if you are using a schedule-40 trap, or you may use a trap adapter. A trap adapter will be needed if the trap is metal and the trap arm is plastic. Trap adapters glue onto pipe just like any other fitting. One end of the adapter is equipped with threads to accept a slip-nut.

> Traps for lavatories can be metal or plastic. Slip nuts are generally employed to make connections. Some plastic traps are glued to trap arms and connect to the drainage tailpiece with a slip nut. Metal traps are secured at both ends with slip nuts.

Start by placing the trap on the tailpiece. To do this, remove the slip nut from the vertical section of the trap. Slide the slip nut onto the tailpiece and follow it with the washer that was under it; the washer may be nylon or rubber. Put the trap on the tailpiece and check the alignment with the trap arm. It may be necessary to use a fitting to offset the trap arm in the direction of the trap.

If the trap is below the trap arm, you will have to shorten the tailpiece. The tailpiece is best cut with a pair of roller cutters, but it can be cut with a hacksaw. You may have to remove the tailpiece to cut it. If the trap is too high, you can use a tailpiece extension to lower it. A tailpiece extension is a tubular section that fits between the trap and the tailpiece. The extension may be plastic or metal, and it is held in place with slip-nuts and washers.

Once the trap is at the proper height, you must determine if the trap arm needs to be cut or extended. Extending the trap arm can be done with a regular coupling and pipe section. If you are using a schedule-40 plastic trap, it is glued onto the trap arm. If you are using a metal trap, the long section of the trap will slip into a trap adapter. You may have to shorten the length of the trap's horizontal section. When using a trap adapter, slide the slip nut and washer on the trap section; then insert the trap section into the adapter and secure it by tightening the slip nut. Once the trap-to-trap-arm connection is complete, tighten the slip nut at the tailpiece.

With the water cut off to the supply pipes, install the cut-offs for the lavatory. Remove the aerator (the piece screwed onto the faucet spout) from the faucet. If you don't remove the aerator before you run water through the faucet for the first time, it will often become blocked with debris and cause an erratic water stream. Other than to test for leaks, your work with the lavatory is done.

TUB AND SHOWER TRIM

Installing tub and shower trim is easy, but connecting a tub waste and overflow is difficult if you don't have any help. Let's take a look at what you need to know to trim out your tub or shower.

Shower Trim

When you want to install shower trim, start with the shower-head. Be sure the main water supply is turned off, and unscrew the stub-out from the shower head ell. Slide the escutcheon that came with the shower assembly over the shower arm. Apply pipe dope or sealant tape to the threads on each end of the shower arm. Screw the showerhead on the short section of the arm where the bend is. Screw the long section of the arm into the threaded ell in the wall.

Use an adjustable wrench on the flats around the shower-head to tighten all connections. If you must use pliers on the arm, keep them close to the wall so that the escutcheon will hide scratch marks.

Now you are ready to trim out the shower valve. How this is done will depend on the type of faucet you roughed in. Follow manufacturer's suggestions. If you installed a single-handle unit, you will normally install a large escutcheon first. These escutcheons normally use a foam gasket, eliminating the need for plumber's putty. Then the handle is installed, and the cover cap is snapped into place over the handle screw.

If you are using a two-handle faucet, you will normally screw chrome collars over the faucet stems. These may be followed by escutcheons, or the escutcheons may be an integral part of the sleeves. Putty should be placed where the escutcheons come into contact with the tub wall. Then the handles are installed.

Tub Faucets

Tub faucets are trimmed out in the same ways as shower faucets. However, you will have a tub spout to install. Some tub spouts slide over a piece of copper tubing and are held in place with a setscrew. Many tub spouts have female-threaded connections either at the inlet or the outlet of the spout. If you are dealing with a threaded connection, you must solder a male adapter onto the stub-out from your tub valve or use a threaded ell and galvanized nipple. The type of spout that slides over the copper and attaches with a setscrew is by far the easiest to install. You should place plumber's putty on the tub spout where it comes into contact with the tub wall.

Tub Wastes

Tub wastes are difficult to install when you are working alone. The tub waste and overflow can take several forms. It may be made of metal or plastic. It can use a trip lever, a push button, a twist-and-turn stopper, or an old-fashioned rubber stopper. The tub waste may go together with slip-nuts or glued joints. Follow the directions that come with your tub waste.

The first step for installing a tub waste is the mounting of the drain. Unscrew the chrome drain from the tub shoe. You will see a thick black washer. Install a ring of putty around the chrome drain and apply pipe dope to the threads. Hold the tub shoe under the tub so that it lines up with the drain hole. Screw the chrome drain into the female threads of the shoe. The black washer should be on the bottom of the tub between the tub and the shoe. Once the chrome drain is hand-tight, leave it alone for now.

The tub shoe has a tubular drainage pipe extending from it. Point this drain towards the head of the tub where the faucets are. Take the tee that came with the tub waste and put it on the drainage tube from the shoe. The long drainage tube that will accept the tub's overflow should be placed in the top of the tee. You want the face of the overflow tube to line up with the overflow hole in the tub. Cut the tubing on the overflow or shoe as needed for a proper fit. The cuts are best made with roller-cutters, but they can be made with a hacksaw.

You should have a sponge gasket in your assortment of parts. This gasket will be placed on the face of the overflow tubing between the back of the tub and the overflow head. From inside the bathtub, install the faceplate for the overflow. For trip-lever styles you will have to fish the trip mechanism down the overflow tubing. For other types of tub wastes you will only have a cover plate to screw on. Tighten the screws until the sponge gasket is compressed.

Now tighten the drain. This can be done by crossing two large screwdrivers and using them between the crossbars of the drain. Turn the drain clockwise until the putty spreads out from under the drain. The last step is connecting the tub waste to the trap. This can be done with trap adapters or glue joints, depending upon the type of tub waste you have.

> If you install a tub waste that has mechanical joints, such as slip nuts, you must provide an access panel to service the tub waste. When the tub waste is soldered brass or glued plastic, an access panel is not needed.

Apply joint compound to the threads of the tailpiece if you're using a metal waste and screw the tailpiece into place. From here on it is just like hooking up a lavatory drain.

INSTALLING ELECTRICAL FIXTURES AND DEVICES

Installing electrical fixtures and devices is quite simple, but caution must be observed due to the risk of electrocution. Never trust a wire until it has been tested with a meter.

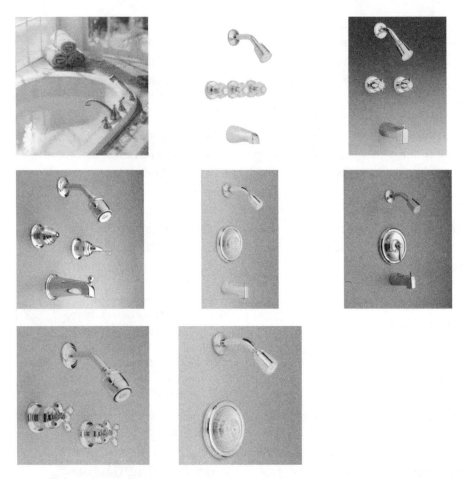

Figure 16.8   A variety of tub, shower, and tub/shower faucets. *Courtesy of Moen, Inc.*

Installing Wiring Devices

When installing wiring devices, there is a color-code system that should be followed. Green wires or bare copper wires should be used as ground wires and attached to green screws. Red wires should be considered hot wires and will normally attach to brass or chrome screws. Black wires are also considered hot and generally attach to brass screws. In many cases

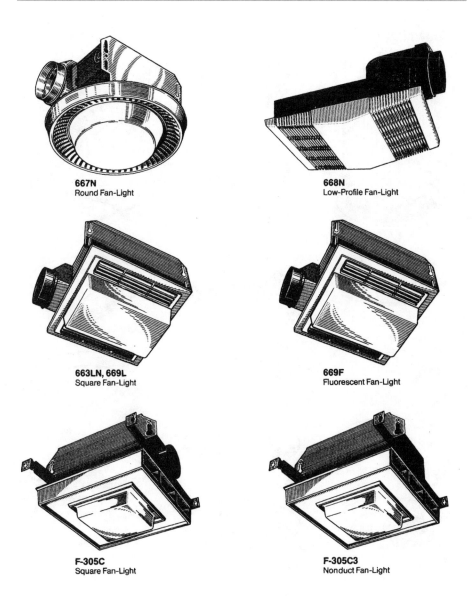

667N
Round Fan-Light

668N
Low-Profile Fan-Light

663LN, 669L
Square Fan-Light

669F
Fluorescent Fan-Light

F-305C
Square Fan-Light

F-305C3
Nonduct Fan-Light

Figure 16.9   A collection of fan-light combinations. *Courtesy of Nutone*

Downlights

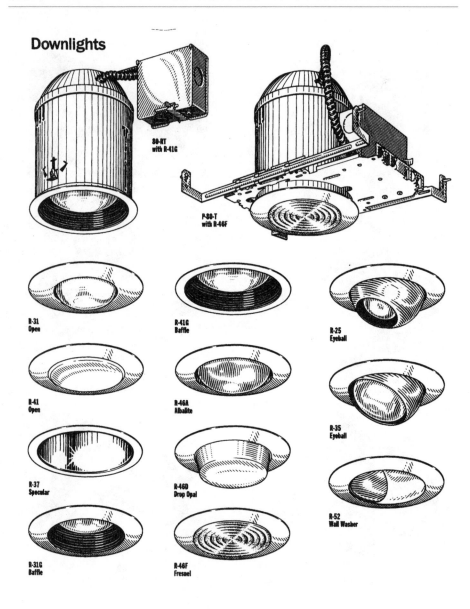

Figure 16.10   Recessed lighting options. *Courtesy of Nutone*

white wires serve as a neutral wire and connect to chrome screws, but they can be hot.

Wall Plates and Switch Covers

Wall plates and switch covers simply mount over outlet boxes and switch boxes. They are held in place with screws.

Installing Light Fixtures

Installing light fixtures is usually a matter of matching up feed wires with fixture wires and mounting the fixture. Most fixtures have threaded studs that hold them to their electrical box. Consult the directions that come with your light fixtures and follow the manufacturer's recommendations.

This has been a long chapter, but you should have learned a great deal. With all this work out of the way, you may feel like you are home free, but don't relax too much. The last few days of a job are when many accidents happen. Let's turn to the next chapter and see how slacking off near the end could cause problems.

17

Finishing the Job

When the time comes for the dust to settle, you will most likely be tired of remodeling and elated with your success. This is when you have to peel the stickers off fixtures, remove the big black letters from vinyl flooring, and pull together other minor details. This is a glorious time for remodelers, homeowners, and professionals alike. However, this is also a time when many avoidable accidents happen and put a damper on the jubilant success of a job well done.

All too often remodelers let their guard down near the end of a job. An anxious remodeler uses a razor blade to remove a sticker from a fiberglass bathtub, and the tub is scratched to a point where fiberglass repair is required. There are plenty of disasters waiting to happen at the end of a job.

Can you avoid last-minute problems with your remodeling project? Yes, but you will have to remain alert and attentive to what you are doing. Let's look at a list of common last-minute details and see how you can avoid problems that will ruin your remodeling experience.

VINYL FLOORS

Here is a list of thing to look out for with vinyl flooring:

- Vinyl floors can be cut or torn when installing cabinets and fixtures. This disaster can be avoided by covering the new flooring with cardboard while the equipment is being installed.

- The bold black letters on vinyl flooring can be removed with standard household cleaners. Warm water with a little floor cleaner and a sponge mop will make the marks disappear.

- If a new vinyl floor develops bubbles in its surface, a straight pin will solve the problem. The pin can be used to pop the bubble and release trapped air.

WALLS

Walls are frequently abused during the final phases of a remodeling job. Touching up the paint on the walls will not take long, but the touch-up should be feathered out to conceal its presence. If you just dab a little paint on a marred surface, the repair will be obvious. Apply a thin coat of paint to a large enough area so that it blends in well. Dings and small holes poked in walls can be filled with drywall compound, sanded, and painted.

Countertops can be cleaned with standard household cleaners, but heavy abrasives should be avoided.

It is not unusual for ceilings to suffer marks and holes from baseboard trim installation. Ceilings can be repaired with the same techniques used on walls.

WINDOWS

New windows are often covered with stickers. These stickers can be removed in short order with a straight-edged razor

The Masters Group, Inc.

PMB # 300 13 Gurnet Road
Brunswick, Maine 04011
207-729-8357 (Phone)
207-798-5070 (Fax)
tmg1@mfx.net (Email)

CERTIFICATE OF COMPLETION
AND ACCEPTANCE

Contractor: _____

Customer: _____

Job name: _____

Job location: _____

Job description: _____

_____

Date of completion: _____

Date of final inspection by customer: _____

Date of code compliance inspection and approval: _____

Defects found in material or workmanship: _____

_____

_____

_____

_____

ACKNOWLEDGMENT

Customer acknowledges the completion of all contracted work and accepts all workmanship and materials as being satisfactory. Upon signing this certificate, the customer releases the contractor from any responsibility for additional work, except warranty work. Warranty work will be performed for a period of _____ from the date of completion. Warranty work will include the repair of any material or workmanship defects occurring between now and the end of the warranty period. All existing workmanship and materials are acceptable to the customer and payment will be made, in full, according to the payment schedule in the contract, between the two parties.

_____   _____
Customer   Date   Contractor   Date

Figure 17.1   Certificate of Completion and Acceptance

blade. The use of a holder for the blade will reduce the likelihood of an accidental cut to your fingers.

PLUMBING FIXTURES

Plumbing fixtures frequently have stickers attached to them. Hot water is the best way to remove them. A razor blade can be used, but take care not to scratch the finish of the fixture.

INSTALL ACCESSORIES

You usually install accessories at the end of the job. In a bathroom-remodeling job the accessories might be towel rings, towel racks, shower curtains, doors, toothbrush holders, soap holders, toilet-paper holders, and so forth. These accessories can cause a lot of grief. If you make a hole in a new wall at the wrong location, it will have to be patched and painted. Don't install accessories on impulse. Plan for them as you plan for the major parts of your job. If you are unable to attach the accessories to wall studs, use screws and expanding anchors to secure them to your drywall. Take the time to put a level on appropriate accessories such as towel racks. Negligence at this stage of the game can be costly.

Light fixtures are not sold with accompanying light bulbs. To avoid having a dark room once your fixtures are installed, remember to buy bulbs. Check the fixtures for wattage ratings, and do not exceed the recommended wattage on the bulbs you install.

CLEANING UP

When you are cleaning up the job, be aware of your new walls, fixtures, windows, and cabinets. A mop handle can do a lot of damage if it is jammed into a window or wall. Even if the handle doesn't knock a hole in your new wall, it can leave a long streak on the finished surface.

Figure 17.2 A lighted medicine cabinet and vertical linen storage unit add to this bathroom. *Courtesy of Quaker Maid*

Figure 17.3 Vertical storage units. *Courtesy of Quaker Maid*

Figure 17.4   An affordable, oak-frame mirror is a good choice for most bathrooms. *Courtesy of Nutone*

Don't use cleaners with abrasive agents on your plumbing fixtures, floors, or countertops. The heavy grit in these cleaners is not needed with new fixtures, and it can damage them.

Remove all debris from your job before you do your touch-up work. Cardboard boxes and leftover lumber can wreak havoc with the paint on new walls.

Don't make the mistake of sweeping or cleaning your room immediately after doing paint touch-up work. The fresh paint will act like a magnet to the dust you stir up.

FINAL INSPECTION

A final inspection is needed before you pay your contractor. Here are some examples of what to look for:

- Do a final inspection on your work both during the day and at night. Flaws that go unnoticed in one type of light will show up in another.

- Check all plumbing connections closely for leaks.

Figure 17.5   A newly remodeled bathroom can bring great satisfaction.
Courtesy of Armstrong

- Test all your electrical outlets and fixtures.
- Go back to your production schedule and note the work you have done. Then check all the work to make sure that it is satisfactory.
- Inspect your work closely and tend to any corrections immediately. If you put off fixing something, it may remain uncorrected for months.

When you are satisfied that the job is complete and satisfactory, notify the local code-enforcement office to make a final inspection. Final inspections often involve different inspectors. You may see a plumbing inspector, an electrical inspector, a heating inspector, and a building inspector. A fire inspector may even visit your job. Get these inspections out of the way and retain the final approvals for your files.

Guess what—you are done! At this point you have completed your job. Now you have the fun of decorating and enjoying your new space. The stress and tension are gone. You made it. Congratulations! Go relax in the comfort of knowing that you did a good job. Don't hesitate to pat yourself on your back. Remodeling is not difficult, but it does take a lot of effort. Average people can do it, and I suspect that you are more than capable of running a successful job by this point, so go forward and make your mark in the remodeling world.

Index